THE WORLD'S CLASS

SIKES

OTH

CHARLES DICKENS wa … outh, where his father was a clerk in th … removed to London in 1815, and in 1817 to Chatha … the happiest years of Dickens's childhood were spent. The … rned to London in 1822, but their fortunes were severely impaired. Dickens was withdrawn from school, and in 1823 sent to work in a blacking-warehouse managed by a relative. His father was imprisoned for debt. Both experiences deeply affected the future novelist. Once his father's financial position improved, however, Dickens returned to school, leaving at the age of fifteen to become in turn a solicitor's clerk, a shorthand reporter in the law courts, and a parliamentary reporter. In 1833 he began contributing stories to newspapers and magazines, later reprinted as *Sketches by 'Boz'*, and in 1836 started the serial publication of *Pickwick Papers*. Before *Pickwick* had completed its run, Dickens, as editor of *Bentley's Miscellany*, had also begun the serialization of *Oliver Twist* (1837–8). In April 1836 he married Catherine Hogarth, who bore him ten children between 1837 and 1852. Finding serial publication both congenial and profitable, Dickens published *Nicholas Nickleby* (1838–9) in monthly parts, and *The Old Curiosity Shop* (1840–1) and *Barnaby Rudge* (1841) in weekly instalments. He visited America in 1842, publishing his observations as *American Notes* on his return and including an extensive American episode in *Martin Chuzzlewit* (1843–4). The first of the five 'Christmas Books', *A Christmas Carol*, appeared in 1843 and the travel-book, *Pictures from Italy*, in 1846. The carefully planned *Dombey and Son* was serialized in 1846–8, to be followed in 1849–50 by Dickens's 'favourite child', the semi-autobiographical *David Copperfield*. Then came *Bleak House* (1852–3), *Hard Times* (1854), and *Little Dorrit* (1855–7). Dickens edited and regularly contributed to the journals *Household Words* (1850–9) and *All the Year Round* (1859–70). A number of essays from the journals were later collected as *Reprinted Pieces* (1858) and *The Uncommercial Traveller* (1861). Dickens had acquired a country house, Gad's Hill near Rochester, in 1856 and he was separated from his wife in 1858. He returned to historical fiction in *A Tale of Two Cities* (1859) and to the use of a first-person narrator in *Great Expectations* (1860–1) both of which were serialized in *All the Year Round*. The last completed novel, *Our Mutual Friend*, was published in 1864–5. *Edwin Drood* was left unfinished at Dickens's death on 9 June 1870.

PHILIP COLLINS is Professor of English at Leicester University and author of *Dickens and Crime*, *Dickens and Education*, *The Impress of the Moving Age*, *A Dickens Bibliography*, and *Reading Aloud: A Victorian Métier*. He is editor of *A Christmas Carol: The Public Reading Version*, and *Dickens: The Critical Heritage*.

THE WORLD'S CLASSICS

CHARLES DICKENS

Sikes and Nancy

AND

Other Public Readings

Edited with an introduction and notes by

PHILIP COLLINS

Oxford New York

OXFORD UNIVERSITY PRESS

1983

Oxford University Press, Walton Street, Oxford OX2 6DP
London Glasgow New York Toronto
Delhi Bombay Calcutta Madras Karachi
Kuala Lumpur Singapore Hong Kong Tokyo
Nairobi Dar es Salaam Cape Town
Melbourne Auckland
and associates in
Beirut Berlin Ibadan Mexico City Nicosia

Text first published 1975 by Oxford University Press
This edition first published 1983 as a World's Classics paperback

British Library Cataloguing in Publication Data
Dickens, Charles, 1812–1870
Sikes and Nancy and other public readings.—
(The World's classics)
I. Title II. Collins, Philip
823'.8[F] PR4553
ISBN 0–19–281617–9

Library of Congress Cataloging in Publication Data
Dickens, Charles, 1812–1870.
Sikes and Nancy and other public readings.
(The World's classics)
Bibliography: p.
Contents: A christmas carol — The chimes —
The story of Little Dombey — [etc.]
1. Recitations. I. Collins, Philip Arthur William.
II. Title. III. Series.
PR4552.C58 1983 823'.8 82–14354
ISBN 0–19–281617–9 (pbk.)

Printed in Great Britain by
Hazell Watson & Viney Limited
Aylesbury, Bucks

CONTENTS

INTRODUCTION

DICKENS 'reads as well as an experienced actor would—he is a surprising man', wrote W. C. Macready, the leader of the English stage—but this was in 1838, when the novelist had been reading him a new play, twenty years before his career as a paid public reader began: and when that happened Thomas Carlyle, a man not easily impressed, judged that Dickens acted 'better than any Macready in the world; a whole tragic, comic, heroic *theatre* visible, performing under one *hat*, and keeping us laughing . . . the whole night.'[1] Macready himself was overwhelmed, too: Dickens's performance in *Sikes and Nancy*, he said, was the equivalent of his playing 'two Macbeths' in his 'best days' and his 'truly artistic' rendering of *David Copperfield* amazed as profoundly as it moved him (below, pp. 136, 230). Dickens had nearly become a professional actor at the age of twenty and never stopped hankering for the footlights. In the years before the Readings he had been a brilliant amateur actor, and he often expressed his joy in commanding a large audience: 'There's nothing in the world equal to seeing the house rise at you, one sea of delightful faces, one hurrah of applause!'[2] Many letters and conversations attest his conviction that he would have been as successful on the stage as in literature (a belief in which some good judges supported him), and at least in some moods he regretted that he had become an author instead of an actor. His public readings of his own works provided a unique opportunity to combine these two great talents and passions.

For some years before he read in public, he had given occasional pre-publication readings of his latest story or serial instalment to his family or friends. After one of these, in 1846, he mused that 'in these days of lecturings and readings'—a phrase to which we must return—'a great deal of money might possibly be made (if it were not infra dig) by one's having Readings of one's own books. It would be an *odd* thing. I think it would take immensely' (*Life*, pp. 424–5).[3] His friend, later his biographer, John Forster, to whom he wrote this, was discouraging, and never ceased to regard the Readings as 'infra dig', and as a declension from Dickens's higher duties as a great author to a lower and more

[1] *Diaries of William Charles Macready 1833–1851*, ed. William Toynbee (1912), i. 480; J. A. Froude, *Thomas Carlyle: a History of his Life in London 1834–1881* (1884), ii. 270.

[2] Mary Cowden Clarke, *Recollections of Writers* (1878), p. 324.

[3] References to frequently-cited texts will be given thus in the text: see References and Abbreviations, below, p. xxiv. Should 'one's having' here be 'one's giving'?

ephemeral art: and no more was heard of this notion for a few years. Then, in 1853, he offered to give some readings to raise funds for an adult-educational establishment in Birmingham. (Adult education was a cause dear to his heart: most of the subsequent unpaid 'charity' readings were given for similar establishments in provincial cities, and many of his published *Speeches* were given at Mechanics' Institutes and the like.) Thus his first public readings were given, after Christmas 1853, in Birmingham Town Hall, *A Christmas Carol* and *The Cricket on the Hearth* being the two items. His triumph there was widely reported, and soon he was inundated with requests to repeat these performances elsewhere. Some institutions offered him a fee but, though tempted, he declined this. The idea of reading for his own profit, however, remained in his mind. Over the next few years he gave a dozen or so more 'charity' readings, always of the *Carol* and usually around Christmas, and he gave a few more such unpaid performances, mostly for the Chatham Mechanics' Institute, after turning professional in 1858.

One reason for his now giving paid readings was that he had recently depleted his finances by buying a country house, Gad's Hill, and he rightly guessed that he could quickly and surely earn more ready cash by reading than by writing. His main reason, however, was that he was feeling very 'unsettled'—less absorbed than before by his writing, depressed about political and social developments, and, above all, deeply unhappy in his marriage, partly because of his recent infatuation with a young actress, Ellen Ternan. As usual when he was 'unsettled', he felt the need for some energetic outlet: the physical and emotional strain—and the excitement—of performance offered him this, while the adoration of his audiences would provide a welcome emotional gratification. Within a fortnight of his professional début on 29 April 1858, his marriage had broken up. That he continued to give Readings intermittently over the next dozen years, long after this emotional crisis, shows of course that he had other reasons for doing so than those impelling this first season. Loving the footlights, he manifestly enjoyed his audiences' affection and admiration. He knew that he was good at reading, and was not so self-denying as to abstain from parading his skills in public. Moreover he was justly proud of what he had written, and loved giving an oral interpretation of his own work which could enhance its meaning, for his audiences, 'a thousandfold', as the more enthusiastic put it. Reviewers reached for similes to describe what a revelation of these familiar texts the Readings provided. To hear Dickens read instead of reading the books oneself was like meeting someone instead of getting a letter from him—or like seeing a stereoscopic instead of a two-dimensional photograph, or a great painting instead of an engraving of it—or it was to experience 'surprises and new thoughts that come like revelations of suddenly discovered racts in the lives of old acquaintances' (*New York Herald*, 13 December 1867). No wonder that Dickens got much satisfaction beyond the financial from the exercise of this talent, and that he kept returning to it.

His original plans, in 1858, had been quite modest: to read the *Carol* four or six times in London, and then to undertake a provincial tour of about forty performances, after which he would return to the more lucrative of these locations 'to read a new Christmas story written for that purpose'. Beyond this, he glimpsed the possibility of earning ten thousand pounds in America 'if I could resolve to go there' (*Life*, p. 647n.). But before his London readings began, he had expanded his repertoire to include other Christmas Books: *The Cricket*, *The Chimes* and—though he never actually performed it—*The Haunted Man*. The four or six London readings became seventeen, and the provincial tour from August to November brought his total to a hundred, soon followed by a Christmas series in London. Meanwhile his repertoire had changed almost completely. By the end of the year, only the *Carol* had survived from his initial programme, and it had been joined by five other items. Moreover the nightly pattern had changed. Instead of a programme consisting of one item lasting two hours, it now and henceforth nearly always consisted of two: a long one, such as the *Carol* (now reduced to eighty or ninety minutes), followed, after a brief interval, by a short—usually a comic—one, comparable to a theatrical 'afterpiece'.

He then became busy writing *A Tale of Two Cities*. He never undertook substantial Readings engagements while composing a novel, and his next big venture was a provincial tour in the autumn and winter of 1861–2, after completing *Great Expectations*. For this he had refurbished his repertoire, devising six new items, though three of these were never performed. Short London seasons followed in 1862 and 1863, and then another long gap while *Our Mutual Friend* was being written. He then undertook a provincial tour (April to June 1866) under the management of the impresarios, Chappells; previously he had hired his own manager, but his trusty manager Arthur Smith had died, and no satisfactory successor had emerged. Chappells provided a manager, George Dolby, who stayed with him to the end, becoming an intimate associate and later writing an excellent memoir of these years with Dickens. A further Chappell tour followed in 1867, and then at last Dickens made his long-contemplated American tour, delayed by the Civil War and other considerations. He undertook a gruelling tour through the American winter (December 1867 to April 1868), at a time when his health was deteriorating rapidly. His financial reward, however, was splendid. American performances netted him almost three times as much as British ones, and he cleared nearly £20,000 profit, an enormous sum for four months' work, at the money-values of the time.

A few additional items had lately entered his repertoire, notably the very popular *Doctor Marigold*, and on his return from America he devised, as a special attraction for his farewell tour of Britain, his most famous and spectacular piece, the horrific *Sikes and Nancy*. This tour began in October 1868 and ended prematurely the following April when he collapsed with strain. Rapidly recuperating, he was determined to give his farewells in London, and his doctors, with doubtful wisdom, allowed him to do so, once or twice a week from January to March 1870. Within three months of his Final Farewell, he was dead, and the lapidary words of his Farewell speech, '. . . from these garish lights I vanish now for evermore', were inscribed on the funeral card distributed at Westminster Abbey.

Altogether he had given about 472 public readings, including 27 for charity and 75 in America. The most popular items (60 or more performances given) were, in this order: *The Trial from 'Pickwick'*, *A Christmas Carol*, *Boots at the Holly-Tree Inn*, *Doctor Marigold*, *David Copperfield*, *Mr. Bob Sawyer's Party* and *Mrs. Gamp*. It must of course be remembered, however, that some items were in repertoire many years longer than others: thus, *Sikes and Nancy* joined the repertoire only three months before Dickens collapsed on his final tour. One other kind of figure is of interest: financial. During his first tour in 1858 he earned over £40 clear a night, more than Macready could command at the peak of his career, and by the Farewell season Chappells had doubled this sum. Jenny Lind, 'the Swedish nightingale', was, I think, the only performer of his period who could rival his rate of earning. From first to last, the Readings were a huge box-office success, and nearly half of the £93,000 Dickens left in his will had come from this activity. George Dolby rightly commented that, on the one hand, these handsome results 'were purchased at the dear cost of the sacrifice of his health', but also that, on the other, 'setting aside his pecuniary profits, the pleasure he derived from [this career] is not to be told in words' (Dolby, p. 451).

These were by no means mere readings-aloud of the text, but elaborate dramatic performances: and similarly what Dickens 'read' (in fact he recited, for he knew his scripts by heart, and often improvised, or improved upon his original version) was by no means the published text of his fictions, but a much amended, abbreviated and sometimes heightened revision of them. When he started, with the Christmas Books, his task was relatively simple, for these were narratives which, with a few omissions, could be given entire, and they had been printed in a large easily-legible type. So he stuck pages of the original edition on to a larger page, which gave more room for marginalia (rewordings, or stage-directions such as 'Tone to Mystery', 'Very Strong' or 'Very Pathetic' and similar reminders), and then, in rehearsal, he cut the text to a manageable length. When he began preparing items from his novels or magazine Christmas numbers, however, the degree of selection and re-arrangement, and the smaller size of the original print, made this procedure impracticable: so he did a quick scissors-and-paste job and sent this stuck-in text to a printer—as a later author might send a manuscript to a typist—to obtain a legible working copy.

His 'prompt-copies' (as I shall call these booklets which he had before him on the platform, though he had and needed no prompter) are fascinating documents, much underlined, annotated and scribbled-over during rehearsal and over the years, as was often the case, that the item remained in his repertoire—for he rarely discarded or replaced his original 'prompt-copy', but made do with his old much-amended one. Sometimes variations in the colour of the ink, or his use of pencil or other such idiosyncrasies, make it possible to trace successive stages of his revision. Almost always the tendency of his revisions is to cut, and to sharpen his effects. Jokes are made more pointed, dialect (such as Mr. Peggotty's) is thickened, numerals are increased. Thus, Trotty Veck, 'over sixty' in *The Chimes*, becomes 'over sixty-eight' in the Reading; books at Dotheboys Hall averaged 'about one to eight learners' in *Nicholas Nickleby* but were now rounded up to 'a dozen'; and the Miss Fezziwiggs' admirers, in the *Carol*, increased from six to ten and the number of couples at their father's ball went from 'three or four and twenty' to 'seven or eight and thirty', according to one reporter.[4] New jokes were made: thus instead of Mr. Winkle's being asked by Mr. Justice Stareleigh, 'What's your Christian name, sir?' as in *Pickwick Papers*, the question becomes more withering: 'Have you any Christian name, sir?' Or jokes were transferred from elsewhere in the narrative: for instance Gampisms and Micawberisms from other chapters in the novels were deftly inserted into the *Mrs. Gamp* and *David Copperfield* scripts. Sometimes Dickens extemporized new jokes and elaborations, and the more effective of these became a standard part of his text without his bothering to write them into his prompt-copies; examples will be found in the annotation to Serjeant Buzfuz's speech and Mr. Squeers's 'practical mode of teaching'.

Deletions were made for various reasons, often of course simply to reduce the length of the story or episode. Irrelevant characters or plot-developments were omitted, and sometimes two characters were conflated. Indications of who was speaking to whom, and how, were generally deleted, as were many descriptions of characters' facial

[4] Rowland Hill, cited in my Facsimile edition of *A Christmas Carol: the Public Reading Version* (New York, 1971), p. 190.

expression, gesture, movement and demeanour: these were now superfluous, because Dickens's voice and acting conveyed the information. One other kind of deletion should be mentioned—passages of social criticism. The Readings which were performed were, it will be noted, not drawn from the later and more socially critical novels; indeed, over half of them were taken, not from the novels, but from the Christmas books and stories (a fact discussed below). But even from the Christmas books, the more significant passages of this kind—Scrooge's vision of the terrible children Ignorance and Want, and Will Fern's big social-protest speeches, for instance—were eventually omitted in the Readings, though audiences had been greatly impressed by Dickens's rendering of these two passages when, at an earlier stage, they had been included.

The Readings texts never reached a definitive condition. So long as they remained in repertoire, they were subject to further abbreviation and improvement, while the mode of performance too remained surprisingly fresh and spontaneous, as many critics noted. Even such a well-tried favourite as the *Carol* was not allowed to ossify: Dickens told a friend that he thought he had greatly improved his presentation of it while in America, and an English admirer who often heard him perform this item recorded many phrasings which do not appear in the prompt-copy and noted, too, how both text and presentation varied from performance to performance, not only in details of wording but also in the inclusion or omission of lengthy passages.[5]

Luckily almost all of Dickens's prompt-copies have survived. After his death they were dispersed through the salerooms, but in recent years most of them were reunited in two collections, the Berg and the Suzannet, now situated in New York and London. (Particulars of these and other collections drawn upon in this edition are given in the Acknowledgements below, and in the headnotes to every item.) These prompt-copies provide the text here printed, and their marginalia and underlinings, together with the many newspaper accounts of the Readings, make it possible to reconstruct Dickens's performances, which, as John Hollingshead remarked, amounted to 'running critical commentaries upon his own works' (*Critic*, 4 September 1858). Two reporters who did not write for publication deserve special mention: W. M. Wright and Rowland Hill. Wright attended many of the Readings in America and, with book on knee, made copious notes on Dickens's gestures, vocal effects, textual alterations, etc. Hill, a Bedford journalist, did the same, even more amply, but on the *Carol* alone. Three other major witnesses are George Dolby, his manager, Charles Kent, an

[5] James T. Fields, *Yesterdays with Authors* (1872), p. 241; Rowland Hill, typescript 'Notes' (see 'References and Abbreviations').

intimate friend, and Kate Field, an American journalist, all of whom wrote illuminating books about the Readings: see 'Further Reading', below, for further particulars. The annotation to the text provides an ample selection of such evidence.

What W. M. Wright had on his knee was *The Readings of Mr. Charles Dickens as Condensed by Himself*, published in individual booklets and later in collected hardback form by Ticknor & Fields (Boston, 1868). Dickens was a close friend of Fields, and had lent him, for this purpose, the prompt-copies of the nine items which he performed in America. This collection, reprinted in London in 1883 and 1907, remained the only one until my edition of all the *Public Readings* (Oxford, 1975), though a few other items had been published separately. In England, Dickens had never published his Readings texts after his initial 1858 season: the venture had proved commercially unsuccessful. The present volume contains Dickens's most popular items, selected on principles explained below, and it includes *Sikes and Nancy*, which was devised after the American tour and had therefore not appeared in the 1868 edition and its reprints.

These Readings constituted 'a wholly unexampled incident in the history of literature', wrote Charles Kent, whom Dickens had permitted to be their official chronicler (Kent, p. 36), and other more objective commentators agreed: 'a new medium for amusing an English audience', 'a novelty in literature and in the annals of "entertainment" ', to quote two reviewers in 1858. But Dickens in 1846, it will be remembered, had referred to 'these days of lecturings and readings', and there were some analogues to his platform career which, while not providing him with a model or indeed rivalling the unique qualities and eminence of his performances, did help to create the conditions in which he could succeed. They are conveniently summarized by a critic during his Farewell tour, who also indicates Dickens's originality:

Hear Dickens, and die: you will never live to hear anything of its kind so good. There has been nothing so perfect, in their way, as those readings ever offered to an English audience. Great actors and actresses—Mrs Siddons herself among them—have read Shakespeare to us; smaller actors, like the Mathews, elder and younger, John Parry, and others, have given 'entertainments' of a half-literary, half-histrionic order; eminent authors, like Coleridge, and Hazlitt, and Sydney Smith, and Thackeray, have read lectures—and many living authors lecture still—but all those appearances, or performances, or whatever else they may be called, are very different from Mr. Dickens' appearances and performances as a reader. He is a story-teller; a prose *improvisatore*; he recites rather than reads; acts rather than lectures. His powers of vocal and facial expression are very great; he has given them conscientious culture; and he applies them heartily and zealously to the due presentment of the creations of his own matchless genius. It has been said than an author is generally either greater or less than his works—that is, that in the works we see the best of a man, or that in the man there is better stuff than he is able to put into his works. In Mr. Dickens, as a reader, each is equal to the other. His works could have no more perfect illustrator; and they are worthy of his best efforts as an artist. (*Scotsman*, 8 December 1868)

Several factors had lately popularized such recitals, entertainments and lectures. Since the 1830s many Mechanics' Institutes, Literary and Philosophical Societies and similar institutions had been established which provided an audience and a venue for such one-man events, given by professionals, amateurs and semi-professionals. Also some relaxation in the Nonconformist conscience was occurring: there were people who would still never enter a theatre—but 'readings' and other such entertainments were a permissible relaxation. (Thus many local press reports of Dickens's readings note specially that clergymen were present, and hence also such bizarre moral compromises as a popular recitalist, the Reverend J. C. M. Bellew, spouting *Hamlet* while a company of actors mimed the action: staged in a hall instead of a theatre, this could be attended by the godly.) A few authors, most notably Thackeray, had preceded Dickens on to the platform and in touring the country, reading their plays or verses, or giving lectures—'All our literati seem inclined to become "oral instructors"', said the *Illustrated London News*, with some exaggeration, a fortnight after Dickens's professional début—and Thackeray, the year before, had indicated another factor favouring such enterprises when he thanked God 'for the railroad that whisks me about' on his lucrative lecture-circuit. For without the railroads, he wrote, he could never have maintained this one-night-stand schedule.[6] The dozen years of Dickens's readings career coincides, indeed, with the peak of the popularity of such entertainments.

He enjoyed, however, evident and unique advantages over all his rivals; indeed he had no rival. Before he began this secondary career, he had for over twenty years been the unrivalled literary darling of the nation, and of America too. No previous author had commanded such affection as well as respect and admiration. When assessing his chances of success before embarking upon the Readings in 1858, he had referred to 'that particular relation (personally affectionate and like no other man's) which subsists between me and the public', and he was right; as one observer remarked, at the end of his performance 'it was not mere applause that followed, but a passionate outburst of love for the man'.[7] Everyone in the 'vast assembly' at his New York début, wrote a critic, was united in wanting to do homage to

[6] *Letters and Private Papers of W. M. Thackeray*, ed. Gordon N. Ray (1946), iv. 22.

[7] *Life*, p. 646; Moncure D. Conway, *Autobiography* (1904), ii. 7.

. . . a man of true and beneficent genius. To see and feel this was to be deeply and inexpressibly thrilled—was to realize, with gladness and gratitude, the profound devotion of true-hearted men and women to a great natural guide and leader. An immense chord of feeling has been touched and sounded by Charles Dickens. In thirty years of literary life . . . he has created immortal works of art . . . and has won in equal measure the homage and the love of his generation. Something of this affectionate feeling was heartily expressed by his audience last night; nor in all that great throng was there a single mind unconscious of the privilege it enjoyed in being able, even so partially, to thank Charles Dickens for all the happiness he has given to the world. It is a better world because of him. (*New York Tribune*, 10 December 1867)

To quote one other such tribute, from an account of a Farewell performance in Dublin: at its conclusion, 'the people stood up and cheered him lustily. It is not that the world knows Mr. Dickens to be merely a great man; but we all know him to be a good man. And, therefore, his reading is not looked upon as a performance, but as a friendly meeting longed for by people to whom he has been kind' (*Freeman's Journal*, 12 January 1867).

The Readings certainly had this special atmosphere of affection and homage. 'Not looked upon as a performance', said that Irish journalist: but in a critique two days later he did look upon them in this respect, and he reported handsomely: 'It can honestly be said that Mr. Dickens is the greatest reader of the greatest writer of the age.' This Introduction began with Macready's and Carlyle's praise for his vigour and versatility as a recitalist, and hundreds of other such assessments could be quoted from observers eminent, obscure or anonymous. When applying for an audition at Covent Garden in 1832 he had claimed to possess 'a strong perception of character and oddity, and a natural power of reproducing in my own person what I [have] observed in others' (*Life*, p. 59), and his later amateur theatricals, like the Readings, showed that he had great talents as a character-actor. After seeing Dickens's performance in his and Wilkie Collins' drama *The Frozen Deep* (1857), Thackeray exclaimed that, if he turned professional actor, he could make £20,000 a year. Commenting on the same performance (which proved to be the last of his amateur theatricals, for the Readings soon afterwards replaced them in his life), the *Saturday Review* described his acting as 'quiet, strong, and effective', *The Times* remarked that 'Where an ordinary artist would look for "points" of effect he looks for "points" of truth', and *The Leader* went so far as to suggest that if professional actors would learn from Dickens it 'might open a new era for the stage'.[8]

[8] Thackeray reported by William Howitt, in Carl Ray Woodring, *Victorian Samplers: William and Mary Howitt* (Lawrence, Kansas, 1952), p. 184; critiques quoted by Robert Louis Brannan in his *Under the Management of Mr. Charles Dickens: his Production of 'The Frozen Deep'* (Ithaca, N.Y., 1966), pp. 80, 82, 74.

If no author had ever enjoyed 'that particular relation (personally affectionate and like no other man's)' with his public, so also no author of comparable magnitude had possessed in addition such outstanding histrionic powers or had so adored the sight of an audience. Moreover no great author except Shakespeare had written works so manifestly destined for widespread popularity. All these qualities combined with the favourable cultural circumstances described above to ensure that the Readings were an immediate, lasting, conspicuous and well-deserved success. The wonder is that Dickens had not embarked upon them earlier. It must be recognized, however, that they not only harmed his health but deprived posterity of the novels he might otherwise have written. During the dozen years of his 'paid' Readings, from 1858 to 1870, he wrote only two shorter novels, *A Tale of Two Cities* and *Great Expectations*, and the full-length *Our Mutual Friend*; the uncompleted *Edwin Drood* was to follow his retirement from the platform. This is a much slighter literary output than for any earlier dozen years of his career, though there were other factors in his personal and creative life that also help to explain this slackening-off: but certainly the Readings took time and energy, satisfied his artistic urge, and, by paying him so well, decreased the financial incentive which had often encouraged him to return to his desk and pen. John Forster, it must be agreed, was not just being stuffy or captious when he opposed Dickens's giving these Readings 'by which [he admitted], *as much as by his books*, the world knew him in his later life' (*Life*, p. 363; italics mine).

We may regret his having devoted his energies to the Readings, but very few who attended them regretted the occasion. He had magnetic stage presence. Everybody was riveted by his eyes—eyes 'like exclamation points', eyes of 'mingled kindness and sharpness', 'a look of keen intelligence about the strong brow and eye—the look of a man who has seen much and is wide awake to see more', eyes 'unlike anything before in our experience; there are no living eyes like them', to quote four American accounts.[9] 'His powers of vocal and facial expression are very great', that *Scotsman* critic said. He had a voice resonant enough to command audiences of three thousand, though some large vocal effects were beyond his powers, and the voice was flexible, expressive and well-modulated, too. 'I had no conception', said Carlyle, 'before hearing Dickens read, of what capacities lie in the human face and voice. No theatre-stage could have had more players than seemed to flit about his face, and all tones were present.'[10] His hands and fingers were even more expressive than face and voice, according to Kate Field ('What Dickens *does* is frequently infinitely better than anything he says, or the

[9] Quoted in *Dickens: Interviews and Recollections*, ed. Philip Collins (1981), ii. 318, 301, 300, 318.

[10] D. A. Wilson, *Carlyle at Threescore-and-Ten, 1853–1865* (1929), p. 505.

way he says it'), and she notes, for instance, how his fingers performed the Sir Roger de Coverley, in the *Carol*, as if his desk were the dancing-floor and every finger were a leg of one of the Fezziwigs (Field, pp. 31, 33). 'In his ever-active hand', another critic remarked, 'an unlimited power of illustration resides. Frequently a mere motion of the hand shed a hitherto undreamt-of meaning upon a whole passage' (*Belfast Newsletter*, 9 January 1869). Yet his art, though consciously skilful and highly polished—the result, said Charles Kent, of 'a scarcely credible amount of forethought and preparation' (Kent, p. 24), with as many as two hundred rehearsals before an item was performed—was not ostentatious. By the platform standards of the day, his was a restrained, dignified and natural performance, avoiding 'claptraps' and those 'distressing physical exercises which make some elocutionists objects of pity rather than of admiration' (*Brighton Gazette*, October 1868). 'Mr. Dickens', an American journalist wrote approvingly, 'does not rant, nor mouth, nor declaim, nor read after the manner of most actors and elocutionists. He talks, he converses; he never forgets that he is a gentleman of the world, relating to perfectly sensible and rational beings a simple story. There is a nameless charm in this naturalness' (*Hartford Daily Courant*, 19 February 1868). He did not pause while his effects could 'tell', nor linger for applause. He began his performance briskly, without drawing attention to himself or his legendary popularity. Though vigorous, dramatic and resourceful, he was generally thought to have displayed good taste in his mode and level of performance, encouraging his audiences to attend to the text more than to admire its teller, and tactfully maintaining the convention of the public reader, who is aware of his audience's presence, while employing the skills of an actor, who usually pretends that he is unaware of it. He 'carefully avoids making his dramatic faculty too prominent', remarked the *New York Tribune* (11 December 1867): 'He does not, except on very rare occasions, act thoroughly *out*; he suggests, and suggests very forcibly; but he leaves to his hearers to supply what he does not feel it necessary to delineate . . . This is just what the very best reading—that is, reading, and not acting—ought to be.' There were of course less favourable verdicts. Dickens had his off-days and his Readings had their weaker moments: and critics, too, could be captious or impercipient. The overwhelming consensus, however, of the hundreds of critics and other attenders whose accounts I have read is one of satisfaction, admiration, wonder and delight.

To see the beloved and inimitable Boz was in itself a pleasure and privilege, as several quotations above aver, and of course he brought to his rendering of his own compositions a special insight and authority, and he had the technical skill to convey his rich understanding of the text. But also his fiction lent—and still lends—itself to such performance uniquely well. Dickens was not the first nor the last public reader of his works; the names of Bransby Williams from a previous generation and of Emlyn Williams from our own may remind us of this: and no classic English novelist has been so constantly presented in other dramatized forms, for stage, cinema, radio and television. His fiction is intrinsically dramatic, his prose highly aural: and his habits of composition neatly demonstrate this. He heard and saw what he created before he wrote it down. The classic anecdote illustrating this is his daughter Mamie's, about an occasion when, unusually, she was allowed to sit in his study while he was at work (on *Hard Times*, as I have argued elsewhere). After cogitating, he would suddenly rush to a mirror, she recalled, in front of which he made 'some extraordinary facial contortions', and this 'facial pantomime' was accompanied by his 'talking rapidly in a low voice', after which he would return to his desk and write.[11] It was, as it were, a private 'reading' as an immediate preliminary to writing. Maybe it was such curious and memorable practices, and not only the general nature of his imagination, that led Forster to remark on his 'power of projecting himself into shapes and suggestions of his fancy . . . What he desired to express he became.' And, discussing Dickens's early ambition to go on the stage, he comments that no one need regret 'how great an actor was in Dickens lost. He took to a higher calling, but it included the lower' (*Life*, pp. 380, 381). That is, his fiction partook—in no pejorative sense—of the theatrical; and in giving Readings from that fiction ('a substitution of lower for higher aims', as Forster regarded it—*Life*, p. 641), he was both adopting a variant of that 'lower calling' to which he had been so drawn, and also demonstrating his long-standing debt, as a writer, to it.

The success of the Readings, then, depended not only upon the glamour of Dickens's personality and his genius as a recitalist but also upon the special qualities of his art: its mode of characterization, its management of episodes, its narrative prose, and the 'friendly' relation established in the fiction between the author and his public. 'What he is doing now', commented the *New York Tribune* (11 December 1867) on the Readings, 'is only the natural outgrowth of what he has been doing all the days of his life, . . . the spontaneous expression of a great nature in the maturity of its genius'. If, however, the Readings did convey and depend upon significant aspects of his genius as a writer, it must also be acknowledged that much that is important in his literary achievement is poorly represented here. None of the novels after *David Copperfield* (1849–50) provided a Reading which entered his repertoire—and it is the novels of this last twenty years of his career that are now regarded as

[11] Mary Dickens, *My Father as I recall him* [1897], p. 47. This and similar anecdotes are reprinted in *Dickens: Interviews and Recollections*, ed. Philip Collins (1981), i. 121, 127; ii. 272.

his finest work. During and after his lifetime, it is worth noting, this judgment was not widely held: the work up to *Copperfield* was generally more admired, and it remains the most popular area of his fiction. The Readings in his repertoire which were derived from the novels come from these earlier and more popular works: two from *Pickwick*, one each from *Oliver Twist*, *Nickleby*, *Chuzzlewit*, *Dombey* and *Copperfield*. Unfortunately Dickens writes or is recorded as saying very little about which of his own books he most admired, or why he selected his Readings from some and not from others. Maybe he felt that audiences would prefer extracts from the good old favourites, maybe he found it easier to extract short self-contained pieces from the earlier and more episodic novels.[12]

The later novels are often described as 'darker' and it is in them that his deeper and more insistent social criticism appears. The Readings conspicuously omit this important element in his achievement. As was noted above, some striking passages of this kind from earlier stories, such as the vision of Ignorance and Want in the *Carol*, were deleted from his Readings texts. Again Dickens is silent about his doing this, let alone about his motives in so doing, but one may surmise that he felt it inappropriate to strike such notes in an evening's entertainment. The novels he was still writing during these years, and his journalistic articles and public speeches, gave him continuing opportunities for such graver business. It is his humour that the Readings most abundantly represent—and this, as Forster said, was 'his leading quality, . . . his highest faculty [as a novelist] . . . He was conscious of this himself' (*Life*, p. 721). His powers of pathos, much more admired then than later, are displayed, eminently in *Little Dombey* but more incidentally in other items. The storm in *Copperfield* exemplified his more sublime achievement, and *Sikes and Nancy* his command of the terrific ('Terror To The End' was his marginal stage-direction as he approached its climax). With a dozen or more characters in many items, he well illustrated his creative as well as his histrionic facility in this kind, while the wit and energy of his narrative voice were evident throughout. So much, though not all, of his wide-ranging literary achievement was evinced in the Readings.

Five of the items here reprinted come, not from the novels but from the Christmas books of the 1840s and the shorter Christmas stories of the 1850s and 1860s: *A Christmas Carol*, *The Chimes*, *The Poor Traveller*, *Boots at the Holly-Tree Inn* and *Doctor Marigold*, and three of these (the

[12] The items from the novels are, however, only relatively 'self-contained'. Not unreasonably, Dickens assumed that his audiences knew the novels, and therefore did not spend time explaining, for instance, who Mr. Pickwick is or why Mrs. Bardell can accuse him of breach of promise, or who the Micawbers are and why they should be visiting David Copperfield.

Carol, *Boots* and *Marigold*) were among the four most frequently performed items. The Readings had begun, it will be remembered, at Christmas 1853 and the initial 1858 repertoire had consisted entirely of Christmas books. Dickens was of course much associated with the Christmas spirit, and this helps to explain his predilection for these pieces (though the later 'Christmas stories' have little or nothing to do with the Christmas season itself). A further and probably a more important reason for his seeking Readings material from this area of his work was that these were almost the only short narratives that he had written (he was not attracted to the new genre of the short-story) and they were thus the easiest to adapt into readable length and form. A consequence of this choice, however, is that the Readings over-represent his minor achievements, for he did not generally write at his best in these shorter pieces, with the great exception of *A Christmas Carol*, which remains among his most characteristic and finest, as well as his most popular, works. Some indeed of the Readings derived from the 'Christmas' writings, such as *The Poor Traveller* and the items not reprinted here, strike me as feeble or laboured, and two of his most popular items from this source, *Boots* and *Marigold*, are now largely forgotten or not much admired by students of Dickens, and may be regarded, in part or whole, as somewhat dubious literary achievements, though the headnotes to these items seek to explain why, in his performance of them, they so appealed to his audiences. Overall, these Christmas-writings items, with of course the exception of the *Carol*, under-represent or misrepresent his genius as an author, however well they gave opportunities to him as a character-actor and narrator.

In confining his Readings to the earlier novels, then, as in relying so heavily upon the *Carol*, Dickens was—whether to please them, or himself, or both—giving his public what he rightly guessed they would most want. 'It is a difficult thing for a man to speak of himself or of his works,' he said, on what has been described as his first triumph in public speaking (in Edinburgh, 1841). 'But', he continued, 'perhaps on this occasion I may, without impropriety, venture to say a word on the spirit in which mine were conceived.' The word he then spoke was modest, and cannot be taken as an adequate comment on what, over the next three decades, he was to create. But it pointed to something central in that achievement, and was a more fitting comment upon the Readings: 'I felt an earnest and humble desire, and shall do till I die, to increase the stock of harmless cheerfulness' (*Speeches*, pp. 9, 14).

NOTE ON THE TEXT

OF THE twenty-one items that Dickens prepared, the present selection reprints the dozen which effectively constituted his repertoire. It excludes the items prepared but never read (*The Haunted Man*, *The Bastille Prisoner*, *Great Expectations*, *Mrs. Lirriper's Lodgings* and *The Signalman*) and those which had less than ten public performances (*The Cricket on the Hearth*, *Mr. Chops, the Dwarf*, *Barbox Brothers* and *The Boy at Mugby*). Anyone curious about these items will find them in my 1975 edition of *The Public Readings*. The text below is reprinted from that edition, which was based upon Dickens's own copies, most of which incorporated his verbal revisions, marginalia, underlinings, etc. Where more than one of his copies has survived, the later is printed, the aim being to present these Readings in—so far as it can be ascertained—their final form. They are printed in the order in which he devised them, and the headnotes offer brief accounts of their composition. For fuller particulars of their history, textual evolution and reception, and of some minor regularizations of the text, the 1975 edition may be consulted.

The following conventions have been used in printing the text:

(1) Words underlined by Dickens in his prompt-copies are printed in *italic*. Where he has underlined words doubly, or trebly or more, this is recorded in the footnotes. When words appear in italic in his privately-printed copies, they appear in italic here and this is mentioned in the footnotes.
(2) Dickens's marginal stage-directions ('Low', 'Action', 'Cheerful narrative', etc.—almost always underlined, often doubly) are recorded in the footnotes in *italic*.
(3) To help the reader to compare the final Readings text with the novel or story from which it was derived, and to know at what point in the evolution of the text some of the major abbreviations were made, two devices have been used:

* an asterisk signifies that a long deletion (of about 100 words or more) had been made *before* the privately printed edition went to press.
† a dagger signifies that a long deletion (of about 100 words or more) was made, by pen or ink, in the prompt-copy.

In the editorial matter, the style 'Chapter I' refers to sections of the Readings texts; the style 'ch. 1' refers to the original novel or story, page-references to which relate to the Oxford Illustrated edition (1947–58).

FURTHER READING

THE THREE books on the Readings by his contemporaries provide much the best evocation of the performances: those by Kate Field (1871), Charles Kent (1872) and George Dolby (1885). See below, p. xxiv, for particulars. Field's is critically the most acute but Kent and Dolby had the advantage of being among Dickens's intimates, and they provide much fascinating 'offstage' information. Raymund Fitzsimons's *The Charles Dickens Show: an Account of his Public Readings 1858–1870* (1970) provides a lively narrative, drawing upon these sources, and various contemporary accounts are reprinted in *Dickens: Interviews and Recollections*, ed. Philip Collins (1981). Studies in the text of the prompt-copies are listed in *Readings*, p. xl (note), and other recent writings are surveyed in *Victorian Fiction: a Second Guide to Research*, ed. George Ford (New York, 1978), pp. 47, 54. What Dickens's prompt-copies looked like may be discovered from the facsimiles of the *Carol*, *Mrs. Gamp* and *Nickleby*. Facsimiles of virgin copies of his privately-printed scripts of *Copperfield* and *Sikes and Nancy* have also been published. For particulars, see the headnotes to these items.

ACKNOWLEDGEMENTS

MY THANKS go to Mr. Christopher C. Dickens, for permission to use unpublished Dickens material, and to the curators or possessors of the prompt-copies here used or cited: the Trustees of the Henry W. and Albert A. Berg Collection, New York Public Library; of the Suzannet Collection, at the Dickens House, 48 Doughty Street, London; of the Gimbel Collection in the Beinecke Library, Yale University; and Mr. Kenyon Law Starling, of Dayton, Ohio. I am also obliged, for permission to use other manuscripts, to the Henry E. Huntington Library, San Marino, California; the Pierpont Morgan Library, New York; the New York Public Library; and the Dickens House, London. Many of the non-metropolitan British press accounts which I use come from a cuttings-book compiled by the late John Greaves (Dickens House), and some of the American ones from a collection in the Beinecke Library, Yale University, to which Dr. John Podeschi kindly drew my attention. *Readings*, pp. xi–xiii, records my thanks—which I here renew—to many other institutions and individuals whose generous help I received. Finally I thank, for typing this book and for so much else, my colleagues and good friends Mrs. Sylvia Garfield, Mrs. Brenda Tracy, Mrs. Pat Taylor and Miss Anne Sowter, and I dedicate to them this result of our joint labours.

REFERENCES AND ABBREVIATIONS

(Place of publication London, unless otherwise stated)

1. TEXTS OF THE READINGS

Berg, *Gimbel*, *Starling*, *Suzannet* — Dickens's own copies, specified by the names of the collections in which they are now located (see Acknowledgements, above)

Readings — Charles Dickens, *The Public Readings*, ed. Philip Collins (Oxford, 1975)

T & F — *The Readings of Mr. Charles Dickens as Condensed by Himself* (Boston, Ticknor & Fields, 1868)

2. OTHER BOOKS AND MANUSCRIPTS

Coutts — *Letters from Charles Dickens to Angela Burdett-Coutts 1841–1865*, ed. Edgar Johnson (1953)

Dolby — George Dolby, *Charles Dickens as I knew him: the Story of the Reading Tours 1866–1870* (1885). Photographic reprint (New York, 1970)

Field — Kate Field, *Pen Photographs of Charles Dickens's Readings* (1871)

Gordan — John D. Gordan, *Reading for Profit: the Other Career of Charles Dickens. An Exhibition from the Berg Collection* (New York, 1958)

Hill — Rowland Hill, typescript 'Notes on Charles Dickens's *Christmas Carol*' [1930], in the Suzannet Collection, the Dickens House, London

Kent — Charles Kent, *Charles Dickens as a Reader* (1872). Photographic reprint, intro. Philip Collins (1971)

Life — John Forster, *The Life of Charles Dickens*, ed. J. W. T. Ley (1928)

N — *The Letters of Charles Dickens* [Nonesuch edition], ed. Walter Dexter (3 vols., 1938)

Speeches — *The Speeches of Charles Dickens*, ed. K. J. Fielding (Oxford, 1960)

Wright — Marginalia by W. M. Wright in his copy of Dickens's *Readings* (*T & F*, bound into 2 vols.), in the Dickens House, London

A CHRONOLOGY OF CHARLES DICKENS

1812	(7 Feb.) Born at Landport, Hants, to John and Elizabeth Dickens
1815–17	London
1817–22	Chatham, Kent; early education
1823–	London
1824	John Dickens in Marshalsea Debtors' Prison; Dickens employed in Warren's blacking-warehouse
1824–7	At Wellington House Academy
1827–8	Employed as solicitors' clerk
?1829–?1831	Shorthand reporter, at Doctors' Commons; on *Mirror of Parliament*; on *True Sun*
1833–4	First stories published in *Monthly Magazine*
1834	(Aug.)–1836 (Nov.) Reporter on *Morning Chronicle*; sketches published, collected as *Sketches by Boz*, two series Feb. and Dec. 1836
1836	(April)–1837 (Nov.) *Pickwick Papers* (monthly)
1836	(2 April) Marries Catherine Hogarth; lives at Furnival's Inn
1837	(Jan.)–1839 (Jan.) Edits *Bentley's Miscellany*; *Oliver Twist* (monthly—published complete Nov. 1838)
1837	(April)–1839 (Dec.) At 48 Doughty Street. Mary Hogarth dies there, May 1837
1838	(April)–1839 (Oct.) *Nicholas Nickleby* (monthly)
1839	(Dec.) Moves to 1 Devonshire Terrace
1840–1	*Master Humphrey's Clock* (weekly), including *The Old Curiosity Shop* and *Barnaby Rudge*; also monthly, April 1840–Nov. 1841
1842	(Jan.–June) In North America. *American Notes* (Oct.)
1843	(Jan.)–1844 (July) *Martin Chuzzlewit* (monthly) (Dec.) *A Christmas Carol*
1844	(July)–1845 (June) Living in Italy (Dec.) *The Chimes*
1845	(Sept.) First performance by the Amateurs; others in 1846–8, 1850–1 (Oct.)–1846 (March) Planning, editing and contributing to *Daily News* (Dec.) *The Cricket on the Hearth*
1846	*(May) Pictures from Italy* (June–Nov.) Living in Switzerland (Oct.)–1848 (April) *Dombey and Son* (monthly)

	(Nov.)–1847 (Feb.) Living in Paris (Dec.) *The Battle of Life*
1847	(Nov.) Miss Coutts's 'Home for Homeless Women' opened
1848	(Dec.) *The Haunted Man*
1849	(May)–1850 (Nov.) *David Copperfield* (monthly)
1850	(March) Starts *Household Words* (weekly), editing and contributing regularly
1851	(Oct.) Moves to Tavistock House
1852	(March)–1853 (Sept.) *Bleak House* (monthly)
1853	(Dec.) First 'charity' Public Readings given
1854	*Hard Times* (weekly)
1855	(Dec.)–1857 (June) *Little Dorrit* (monthly)
1856	(March) Buys Gad's Hill Place, Kent
1858	(April) Begins paid Public Readings (May) Separates from Mrs. Dickens
1859	(April–Nov.) *A Tale of Two Cities* (weekly and monthly) (May) *All the Year Round* begins (June) *Household Words* ends
1860	*The Uncommercial Traveller* (Oct.) Final removal to Gad's Hill (Dec.)–1861 (Aug.) *Great Expectations* (weekly)
1864	(May–1865 (Nov.) *Our Mutual Friend* (monthly)
1867	(Nov.)–1868 (April) Public reading tour in USA
1869	(April) Breakdown in provincial reading tour
1870	(Jan.–March) Farewell season of Public readings in London (April–Sept.) *The Mystery of Edwin Drood* (monthly; unfinished) (9 June) Dies at Gad's Hill

A CHRISTMAS CAROL

A Christmas Carol, the first public Reading that Dickens ever gave, remained in repertoire throughout his career and was the main item in his final Farewell performance in 1870. This was indeed the quintessential Dickens Reading—as the story which it told has been, ever since its publication in 1843, one of the most central and beloved of his achievements. The first and far the best of the Christmas Books, it is also the most dramatic, a perennial favourite both for recitalists and in adaptations for stage, screen, radio and television. Eminently through the *Carol*, Dickens had identified himself with the Christmas spirit—an identification clearly signalled by his choosing late December for what proved to be the initiation of his public readings career, his three charity performances in Birmingham. The *Carol* was the obvious choice, and it was given on 27 and 30 December 1853; on an intervening night, *The Cricket on the Hearth* was performed. Not only had they a seasonal appropriateness, but also they could without much difficulty be reduced to the length of an evening's entertainment. He had indeed already given a semi-public reading of the *Carol* in 1845, to English residents in Genoa, where he and his family were then living.

Most of the charity performances which Dickens gave between 1853 and 1858 took place around Christmas, and (except for the single performance of *The Cricket* in Birmingham) the programme always consisted of the *Carol.* His attachment to the Christmas spirit appeared further when he was planning his first professional season. Since this began in April 1858, Christmas writings had no seasonal relevance; nevertheless, his original plan was to read nothing that season except Christmas Books. After six weeks he began to perform items drawn from other areas of his work, and soon the *Carol* was the only Christmas Book retained in the repertoire. Over the years, it was included in well over a quarter of his performances, and, with *The Trial from 'Pickwick'*, it constituted his opening performance in every city during his American tour, and for other such notable occasions as his final performances in Paris, New York, and London.

In 1853 the performance took three hours. By May 1858 it was reduced to his normal two-hour length, and by the end of that year to eighty or ninety minutes to accommodate *The Trial* as an afterpiece. This complex abbreviation and revision can be followed in the facsimile of the *Berg* promp-copy (ed. Philip Collins, New York Public Library, 1971). Eventually many episodes were omitted or much reduced. Significantly the only episode left entire was the Cratchits' Christmas dinner—one of the highspots of the Reading, the others being Scrooge's nephew's party culminating in 'It's your Uncle Scro-o-o-o-oge!', Scrooge's dialogue with the boy about the turkey, and '. . . to Tiny Tim, who did NOT die . . .'

'The success was most wonderful and prodigious—perfectly overwhelming and astounding altogether.' Dickens reported after his first performance (*N*, ii. 536). Triumphs followed so regularly during his charity readings, that he had gained a 'gigantic' but thoroughly deserved reputation before turning professional, reported *The Times* (16 April 1858). Of his first London performance, for charity, it was written:

We have rarely witnessed or shared an evening of such genuine enjoyment, and never before remember to have seen a crowded assembly of three thousand people hanging for upwards of two hours on the lips of a single reader . . . Every fragment of the dialogue was treated dramatically—the rendering of each character being equally successful . . . At the close there was an outburst, not so much of applause as of downright hurrahing, from every part [of the hall]. (*Leader*, 4 July 1857)

To quote one other account and tribute, from much later in his Readings career:

His power of facial expression is wonderful; it is as much what he does as what he says that constitutes the charm of his performance. He gives a distinct voice to each character, and to an extraordinary extent assumes the personality of each. At one moment he is savage old Scrooge, at the next, his jolly nephew, and in the twinkling of an eye little timid, lisping Bob Cratchit appears. All this is effected by the play of features as well as the varying tones of voice. It is the comical or the savage twist of the mouth—the former to the right, the latter to the left—the elongation of the face, the roll or twinkle of the eyes, and above all the wonderful lift of the eyebrows, that produce such surprising and delightful effects. And then he not only personates his characters, he performs their actions. This he does by means of wonderfully flexible fingers, which he converts at pleasure into a company of dancers, and makes to act and speak in a hundred ways. He rubs and pats his hands, he flourishes all his fingers, he shakes them, he makes them equal to a whole stage company in the performance of the parts. But then the man himself is also there. Dickens, the author, comes in at intervals to enjoy his own fun; you see him in the twinkle of the eye and the curve of the mouth. When the audience laughs he beams all over with radiant appreciation of the fun. (*Portland [Maine] Transcript*, 4 February 1868)

The annotation below describes some of these visual and vocal effects. One account much cited here was written, astoundingly, sixty years after Dickens's death by a Bedford journalist, Rowland Hill. W. M. Wright, the North American book-on-the-knee recorder, also provides a very full account. Dickens's prompt-copy contains many marginal stage-directions such as 'Cheerful Narrative', but much less underlining than later became his practice.

A Christmas Carol

In Four Staves

STAVE ONE

Marley's Ghost

MARLEY was dead: to begin with. There is no doubt whatever about that. The register of his burial was signed by the clergyman, the clerk, the undertaker, and the chief mourner. *Scrooge*[1] signed it: and Scrooge's name was good upon 'Change, for anything he chose to put his hand to. Old Marley was as dead as a door-nail.

Scrooge knew he was dead? Of course he did. How could it be otherwise? Scrooge and he were partners for I don't know how many years. Scrooge was his sole executor, his sole administrator, his sole assign, his sole residuary legatee, his sole friend, his sole mourner. †

Scrooge never painted out old Marley's name, however. There it yet stood, years afterwards, above the warehouse door: Scrooge and Marley. The firm was known as Scrooge and Marley. Sometimes people new to the business called Scrooge Scrooge, and sometimes Marley. He answered to both names. It was all the same to him.

Oh![2] But he was a tight-fisted hand at the grindstone, was Scrooge! a squeezing, wrenching, grasping, scraping, clutching, covetous old sinner![3]†

Nobody ever stopped him in the street to say, with gladsome looks, 'My dear Scrooge, how are you? when will you come to see me?' No beggars implored him to bestow a trifle, no children asked him what it was o'clock, no man or woman ever once in all his life inquired the way to such and such a place, of Scrooge. Even the blindmen's dogs appeared to know him; and when they saw him coming on, would tug their owners into doorways and up courts; and then would wag their tails as though they said, 'no eye at all is better than an evil eye, dark master!'

But what did Scrooge care!

[1] *Scrooge* here, and in the next paragraph, doubly underlined. Dickens's way of saying his name instantly and vividly conjured up 'Scrooge in the flesh' (Kent, p. 95). 'Old Marley . . . ' was spoken 'Low' (Wright).

[2] 'The *Oh*! was drawn out for some seconds, 3 or 4 perhaps' (Hill).

[3] The ensuing description of Scrooge (in *1843*) was deleted because, in performance, 'we saw and heard it without any necessity for its being explained' (Kent, p. 95). but *T & F* prints the paragraph beginning 'External heat and cold . . . ', which was subsequently deleted. But Kate Field and Hill describe how he read that paragraph, so it was sometimes restored.

[1]Once upon a time—of all the good days in the year, upon a Christmas Eve—old Scrooge sat busy in his counting house. It was cold, bleak, biting, foggy weather, and the city clocks had only just gone three, but it was quite dark already. The door of Scrooge's counting-house was open that he might keep his eye upon his *clerk*,[2] who in a dismal little cell beyond—a sort of tank—was copying letters. Scrooge had a very small fire, but the clerk's fire was so very much smaller that it looked like one coal. But he couldn't replenish it, for Scrooge kept the coal-box in his own room; and so surely as the clerk came in with the shovel, the master predicted that it would be necessary for them to part. Wherefore the clerk put on his white comforter, and tried to warm himself at the candle; in which effort, not being a man of a strong imagination, he failed.[3]

[4]'A merry Christmas, uncle! God save you!' cried a cheerful voice. It was the voice of Scrooge's nephew, who came upon him so quickly that this was the first intimation Scrooge had of his approach.

'Bah!' said Scrooge, 'Humbug!'

'Christmas a humbug, uncle! You don't mean that, I am sure.'

'I do. † *Out*[5] upon merry Christmas! What's Christmas time to you but a time for paying bills without money; a time for finding yourself a year older, and not an hour richer; a time for balancing your books and having every item in 'em through a round dozen of months presented dead against you? If I had my will, every idiot who goes about with "Merry Christmas" on his lips, should be boiled with his own pudding, and buried with a stake of holly through his heart.[6] He should!'

'Uncle!'

'Nephew, keep Christmas in your own way, and let me keep it in mine.'

'Keep it! But you don't keep it.'

'Let me leave it alone, then. Much good may it do you! Much good it has ever done you!'

'There are many things from which I might have derived good, by which I have not profited, I dare say: Christmas among the rest. But I am sure I have always thought of Christmas time, when it has come round—apart from the veneration due to its sacred origin, if anything belonging to it *can* be apart from that—as a good time: a kind, forgiving, charitable, pleasant time: the only time I know of, in the long calendar of the year, when men and women seem by one consent to open their shut-up hearts

[1] Marginal stage direction *Narrative* – 'changing his tone suddenly to a rich mellow note, splendidly inflected' (Hill).

[2] *clerk* doubly underlined.

[3] Dickens's enactment of Bob Cratchit's trying to warm his hands over the candle – reinforced by his inserting 'decidedly' before 'failed' – finally established his command of his audience (Hill; Field, p. 29).

[4] Marginal stage-direction *Cheerful.* The ensuing description of the nephew was deleted; he 'was visibly before us, without a word being uttered' (Kent, p. 99).

[5] *Out* doubly underlined.

[6] This sentence was emphasized 'with a good bang on his reading table' (Hill).

freely, and to think of people below them as if they really were fellow-travellers to the grave, and not another race of creatures bound on other journeys.[1] And therefore, uncle, though it has never put a scrap of gold or silver in my pocket, I believe that it *has*[2] done me good, and *will* do me good; and I say, God bless it!'

The clerk in the tank involuntarily applauded.

'Let me hear another sound from *you*,[3]' said Scrooge, 'and you'll keep your Christmas by losing your situation. You're quite a powerful speaker, sir,' he added, turning to his nephew. 'I wonder you don't go into Parliament.'

'Don't be angry, uncle. Come! Dine with us[4] to-morrow.'

Scrooge said that he would see him—yes, indeed he did. He went the whole length of the expression, and said that he would see him in that extremity first.

'But why?' cried Scrooge's nephew. 'Why?'

'Why did you get married?'

'Because I fell in love.'

'Because you fell in love!' growled Scrooge, as if that were the only one thing in the world more ridiculous than a merry Christmas. 'Good afternoon!'

'Nay, uncle, but you never came to see me before that happened. Why give it as a reason for not coming now?'

'Good afternoon.'

'I want nothing from you; I ask nothing of you; why cannot we be friends?'

'Good afternoon.'

'I am sorry, with all my heart, to find you so resolute. We have never had any quarrel, to which I have been a party. But I have made the trial in homage to Christmas, and I'll keep my Christmas humour to the last. So A Merry Christmas, uncle!'

'Good afternoon!'

'And A Happy New Year!'

'Good afternoon!'[5]

His nephew left the room without an angry word, notwithstanding. The clerk, in letting Scrooge's nephew out, had let two other people in. They were portly gentlemen, pleasant to behold, and now stood, with their hats off, in Scrooge's office. They had books and papers in their hands, and bowed to him.

[1] Against this sentence, Wright notes: 'Moving eyebrows'.
[2] *has* and *will* italic in *1843*.
[3] 'Pointing with right forefinger' at Bob (Wright); *you* italic in *1843*.
[4] Dickens substituted 'with me and my young wife' (Hill).
[5] The successive 'Good afternoons!' spoken crossly and 'with irresistibly humorous iteration' (Wright; Kent, p. 101).

'Scrooge and Marley's I believe,' said one of the gentlemen, referring to his list.[1] 'Have I the pleasure of addressing Mr. Scrooge, or Mr. Marley?'

'Mr. Marley has been dead these seven years. He died seven years ago, this very night.'[2]

'At this festive season of the year, Mr. Scrooge,' said the gentleman, taking up a pen, 'it is more than usually desirable that we should make some slight provision for the Poor and destitute, who suffer greatly at the present time. Many thousands are in want of common necessaries; hundreds of thousands are in want of common comforts, sir.'

'Are there no prisons?'

'Plenty of prisons. † But under the impression that they scarcely furnish Christian cheer of mind or body to the unoffending multitude, a few of us are endeavouring to raise a fund to buy the Poor some meat and drink, and means of warmth. We choose this time, because it is a time, of all others, when Want is keenly felt, and Abundance rejoices. What shall I put you down for?'

'Nothing!'

'You wish to be anonymous?'

'I wish to be left alone. Since you ask me what I wish, gentlemen, that is my answer. I don't make merry myself at Christmas and I can't afford to make idle people merry. I help to support the prisons and the work-houses—they cost enough—and those who are badly off must go there.'

'Many can't go there; and many would rather die.'

'If they would rather die, they had better do it, and decrease the surplus population.' †

At length the hour of shutting up the counting-house arrived. With an ill-will Scrooge, dismounting from his stool, tacitly admitted the fact to the expectant clerk in the Tank, who instantly snuffed his candle out, and put on his hat.

'You'll want all day to-morrow, I suppose?'

'If quite convenient, Sir.'[3]

'It's not convenient, and it's not fair. If I was to stop half-a-crown for it, you'd think yourself mightily ill used, I'll be bound?'

[1] Speaking with just the conciliatory voice 'in which gentlemen-beggars deliver their errands of charity'; Dickens *became* the portly gentleman (Field, p. 29).

[2] Dickens interpolated, at the beginning of this speech, 'Well, you haven't the pleasure of addressing Mr. Marley'. After '. . . these seven years,' he paused, 'opened his eyes widely, and also his mouth, and said, with tragic pauses—"He died—seven years ago—this—very—night! He died today!"' (Hill).

[3] 'When Bob Cratchit lisped [this] out in his timid, trembling tones, . . . the audience caught sight at once of the little, round-faced, deferential, simple-hearted clerk as if he had entered bodily' (*New York Times*, 10 December 1867)—one of many such encomia on this impersonation; cf. Field, p. 30; Kent, p. 101; *Critic*, 4 September 1858, p. 537. 'Sort of lisp in a high frightened timid voice' (Wright).

'Yes, Sir.'

'And yet you don't think *me*[1] ill-used, when I pay a day's wages for no work.'

'It's only once a year, Sir.'

'A poor excuse for picking a man's pocket every twenty-fifth of December! But I suppose you must have the whole day. Be here all the earlier next[2] morning!'

The clerk promised that he would; and Scrooge walked out with a growl. The office was closed in a twinkling, and the clerk, with the long ends of his white comforter dangling below his waist (for he boasted no great-coat), went down a slide, at the end of a lane of boys, twenty times, in honour of its being Christmas-eve, and then ran home as hard as he could pelt, to play at blindman's-buff.

[3]Scrooge took his melancholy dinner in his usual melancholy tavern; and having read all the newspapers, and beguiled the rest of the evening with his banker's-book, went home to bed. He lived in chambers which had once belonged to his deceased partner. They were a gloomy suite of rooms, in a lowering pile of building up a yard. The building was old enough now, and dreary enough, for nobody lived in it but Scrooge, the other rooms being all let out as offices.

Now, it is a fact, that there was nothing at all particular about the knocker on the door of this house, except that it was very large. Also, that Scrooge had seen it night and morning during his whole residence in that place; also that Scrooge had as little of what is called fancy about him as any man in the City of London. And yet Scrooge, having his key in the lock of the door, saw in the knocker, without its undergoing any intermediate process of change: not a knocker, but Marley's face.

Marley's face. With a dismal light about it, like a bad lobster in a dark cellar. It was not angry or ferocious, but it looked at Scrooge as Marley used to look: with ghostly spectacles turned up upon its ghostly forehead.

As Scrooge looked fixedly at this phenomenon, it was a knocker again. † He said 'Pooh, pooh!' and closed the door with a bang.

The sound resounded through the house like thunder. Every room above, and every cask in the wine-merchant's cellars below, appeared to have a separate peal of echoes of its own. Scrooge was not a man to be frightened by echoes. He fastened the door, and walked across the hall, and up the stairs. Slowly too: trimming his candle as he went. †

[1] *me* italic in *1843*.

[2] 'next' italic in *T & F*: a mistake made because a blot from the facing page in *Berg* looked like an underlining. This is one simple proof that *T & F* was set from the prompt-copies.

[3] Marginal stage-direction *Tone to Mystery*. 'Comically' at '... banker's book' (Wright).

[1]Up Scrooge went, not caring a button for its being very dark: darkness is cheap, and Scrooge liked it. But before he shut his heavy door, he walked through his rooms to see that all was right. He had just enough recollection of the face to desire to do that.

Sitting-room, bed-room, lumber-room. All as they should be. Nobody under the table, nobody under the sofa; a small fire in the grate; spoon and basin ready; and the little saucepan of gruel (Scrooge had a cold in his head) upon the hob. Nobody under the bed; nobody in the closet; nobody in his dressing-gown, which was hanging up in a suspicious attitude against the wall.

Quite satisfied, he closed his door, and locked himself in; double-locked himself in, which was not his custom. Thus secured against surprise, he took off his cravat, put on his dressing-gown and slippers, and his night-cap; and sat down before the very low fire to take his gruel. †

As he threw his head back in the chair, his glance happened to rest upon a bell, a disused bell, that hung in the room, and communicated for some purpose now forgotten with a chamber in the highest story of the building.[2] It was with great astonishment, and with a strange, inexplicable dread, that as he looked, he saw this bell begin to swing. Soon it rang out loudly, and so did every bell in the house. This was succeeded by a clanking noise, deep down below; as if some person were dragging a heavy chain over the casks in the wine-merchant's cellar. Then he heard the noise much louder, on the floors below; then coming up the stairs; then coming straight towards his door. It came on through the heavy door, and a spectre passed into the room before his eyes. And upon its coming in, the dying flame leaped up, as though it cried '*I know him! Marley's ghost!*'[3]

The same face: the very same. Marley in his pig-tail, usual waistcoat, tights, and boots. His body was transparent: so that Scrooge, observing him, and looking through his waistcoat, could see the two buttons on his coat behind.[4]

Scrooge had often heard it said that Marley had no bowels, but he had never believed it until now.[5]

[1] The next two paragraphs were deleted, but *stetted* in pencil (which, in this prompt-copy, is usually a sign of a late stage of revision). *T & F* prints the paragraphs, but Wright deletes them.

[2] 'Dramatic. Hand outstretched in the air' (Wright).

[3] A much-praised moment in the Reading: a 'startling effect' as Dickens's voice 'rose to a hurried outcry' (Kent, p. 102). The 'unexpected wild vehemence and weirdness' of the dying flame's cry was 'striking in the extreme' (David Christie Murray, *Recollections* (1908), p. 50).

[4] Another much-praised moment: its 'dismal facetiousness . . . made an abundance of mirth without interrupting the spectral illusion' (*Aberdeen Journal*, 6 October 1858).

[5] 'Low voice, shaking forefinger' (Wright). It was 'said with such a merry twinkle, that the audience always roared with laughter, and sometimes received it with cheers' (Hill).

No, nor did he believe it even now. Though he looked the phantom through and through, and saw it standing before him; though he felt the chilling influence of its death-cold eyes; and noticed the very texture of the folded kerchief bound about its head and chin; he was still incredulous.

'How now!' said Scrooge, caustic and cold as ever. 'What do you want with me?'

'Much!'—Marley's voice, no doubt about it.

'Who are you?'

'Ask me who I *was*.'[1]

'Who *were* you then?'

'In life I was your partner, Jacob Marley.'

'Can you—can you sit down?'

'I can.'

'Do it then.'

Scrooge asked the question, because he didn't know whether a ghost so transparent might find himself in a condition to take a chair. But the ghost sat down on the opposite side of the fireplace, as if he were quite used to it.

'You don't believe in me.'

'I don't.'

'What evidence would you have of my reality beyond that of your senses?'

'I don't know.'

'Why do you doubt your senses?'

'Because a little thing affects them. A slight disorder of the stomach makes them cheats. You may be an undigested bit of beef, a blot of mustard, a crumb of cheese, a fragment of an underdone potato. There's more of gravy than of grave about you, whatever you are!'

Scrooge was not much in the habit of cracking jokes, nor did he feel, in his heart, by any means waggish then. The truth is, that he tried to be smart, as a means of distracting his own attention, and keeping down his horror. †

But how much greater was his horror, when the phantom taking off the bandage round its head, as if it were too warm to wear in-doors, its lower jaw dropped down upon its breast!

'Mercy! Dreadful apparition, why do you trouble me? Why do spirits walk the earth, and why do they come to me?'

'It is required of every man that the spirit within him should walk abroad among his fellow-men, and travel far and wide; and if that spirit goes not forth in life, it is condemned to do so after death.[2] † I cannot tell

[1] *was* and *were* italic in *1843*.

[2] 'In all passages where pathos was the main characteristic, the lines ran easily to rhythm, and the reciter's voice inflected them to almost like blank verse, as when Marley says to Scrooge, "And if that spirit...[*etc.*]"' (*Springfield* [Mass.] *Semi-Weekly Republican*, 21 March 1868).

you all I would. A very little more, is permitted to me. I cannot rest, I cannot stay, I cannot linger anywhere. My spirit never walked beyond our counting house—mark me!—in life my spirit never roved beyond the narrow limits of our money-changing hole; and weary journeys lie before me!'

'Seven years dead. And travelling all the time? You travel fast?'

'On the wings of the wind.'

'You might have got over a great quantity of ground in seven years.'

[1]'Oh! blind man, blind man, not to know, that ages of incessant labour by immortal creatures, for this earth must pass into eternity before the good of which it is susceptible is all developed. Not to know that any Christian spirit working kindly in its little sphere, *whatever it may be*, will find its mortal life too short for its vast means of usefulness. Not to know that no space of regret can make amends for one life's opportunities misused! Yet I was like this man! I once was like this man!'

'But you were always a good man of business, Jacob,' faltered Scrooge, who now began to apply this to himself.[2]

'Business! Mankind was my business. The common welfare was my business; charity, mercy, forbearance, benevolence, were, all, my business. The dealings of my trade were but a drop of water in the comprehensive Ocean of my business!'

It held up its chain at arm's length, as if that were the cause of its unavailing grief, and flung it heavily upon the ground again.

'At this time of the rolling year, I suffer most. Why did I walk through crowds of fellow-beings with my eyes turned down, and never raise them to that blessed Star which led the Wise Men to a poor abode? Were there no poor homes to which its light would have conducted *me*?'[3]

Scrooge was very much dismayed to hear the spectre going on at this rate, and began to quake exceedingly.

'Hear me! My time is nearly gone.'

'I will. But don't be hard upon me! Don't be flowery, Jacob! Pray!'

'I am here to-night to warn you, that you have yet a chance and hope of escaping my fate. A chance and hope of my procuring, Ebenezer.'

'You were always a good friend to me. Thank'ee!'

'You will be haunted by Three Spirits.'

'Is that the chance and hope you mentioned, Jacob? I—I think I'd rather not.'

[1] 'Hands upheld. Fingers outstretched' (Wright).

[2] The extent of Dickens's abbreviation hereabouts is uncertain. In *Berg*, five successive paragraphs have been deleted, at different times (from 'But you were always . . .' to '. . . quake exceedingly'). Later Dickens wrote *Stet* in the margin but (as often when there are several layers of deletion) it is unclear how much is *stetted*. *T & F* omits from 'It held up' to 'conducted *me*'; Hill records that Dickens sometimes included, and sometimes omitted these, and the following, paragraphs. The present text restores all five paragraphs, minus internal cuts.

[3] *me* italic in *1843*. This speech contains four blank verse lines.

'Without their visits, you cannot hope to shun the path I tread.[1] Expect the first to-morrow night, when the bell tolls One. Expect the second on the next night at the same hour. The third upon the next night when the last stroke of Twelve has ceased to vibrate. Look to see me no more; and look that, for your own sake, you remember what has passed between us!'

It walked backward from him; and at every step it took, the window raised itself a little, so that when the Apparition reached it, it was wide open: and [it] floated out through the self-opened window into the bleak dark night.[2] †

Scrooge closed the window, and examined the door by which the Ghost had entered. It was double-locked, as he had locked it with his own hands, and the bolts were undisturbed. Scrooge tried to say 'Humbug!' but stopped at the first syllable. And being—from the emotion he had undergone, or the fatigues of the day, or his glimpse of the Invisible World, or the dull conversation of the Ghost, or the lateness of the hour—much in need of repose; he went straight to bed, without undressing, and fell asleep on the instant.

STAVE TWO

The First of the Three Spirits

When Scrooge awoke, it was so dark, that looking out of bed, he could scarcely distinguish the transparent window from the opaque walls of his chamber † until suddenly the church-clock tolled a deep dull hollow melancholy *One*.[3] † Light flashed up in the room upon the instant, and the curtains of his bed were drawn.

By a strange figure—like a child: yet not so like a child as like an old man, viewed through some supernatural medium, which gave him the appearance of having receded from the view, and being diminished to a child's proportions. Its hair, which hung about its neck and down its back, was white as if with age; and yet the face had not a wrinkle in it, and the tenderest bloom was on the skin. It held a branch of fresh green holly in its hand; and, in singular contradiction of that wintry emblem, had its dress trimmed with summer flowers. But the strangest thing about it was, that from the crown of its head there sprung a bright clear jet of light, by which all this was visible; and which was doubtless the occasion of its using, in its duller moments, a great extinguisher for a cap, which it now held under its arm. †

'Are you the Spirit, sir, whose coming was foretold to me?'

[1] 'Dramatic. Forefinger'; at 'remember . . .' (below) 'Both hands' raised (Wright).

[2] Dickens's revisions hereabouts are inconsistent; he probably omitted, in performance, some of the words after 'it was wide open'. *T & F* falters, providing no exit for Marley's Ghost.

[3] *One* (in marginal manuscript: One in *1843*) underlined three times.

'I am!'

'Who, and what are you?'

'I am the Ghost of Christmas Past.'

'Long past?'

'No. Your past. † The things that you will see with me, are Shadows of the things that have been; they will have no consciousness of us. Rise! And walk with me!' †

It would have been in vain for Scrooge to plead that the weather and the hour were not adapted to pedestrian purposes; that bed was warm, and the thermometer a long way below freezing; that he was clad but lightly in his slippers, dressing-gown, and nightcap; and that he had a cold upon him at that time. The grasp, though gentle as a woman's hand, was not to be resisted. He rose: but finding that the Spirit made towards the window, clasped its robe in supplication.

'I am a mortal, and liable to fall.'[1]

'Bear but a touch of my hand *there*,'[2] said the Spirit, laying it upon his heart, 'and you shall be upheld in more than this!'

As the words were spoken, they passed through the wall, and stood[3] † in the busy thoroughfares of a city. It was made plain enough by the dressing of the shops, that here too it was Christmas-Time.

The Ghost stopped at a certain warehouse door, and asked Scrooge if he knew it.

'Know it! Was I apprenticed here!'[4]

They went in. At sight of an old gentleman in a Welch wig, sitting behind such a high desk, that if he had been two inches taller he must have knocked his head against the ceiling, Scrooge cried in great excitement:

'Why, it's old Fezziwig! Bless his heart; it's Fezziwig alive again!'

Old Fezziwig laid down his pen, and looked up at the clock, which pointed to the hour of seven. He rubbed his hands; adjusted his capacious waistcoat; laughed all over himself, from his shoes to his organ of benevolence; and called out in a comfortable, oily, rich, fat, jovial voice:

'Yo ho, there! Ebenezer! Dick!'

A living and moving Picture of Scrooge's former self,[5] a young man, came briskly in, accompanied by his fellow-'prentice.

'Dick Wilkins, to be sure!' said Scrooge to the Ghost. 'My old fellow

[1] Elaborated to 'I beg your pardon, but being your mortal partner, you don't consider that I am liable to fall down five pairs of stairs!' (Hill).

[2] *there* italic in *1843*. This speech 'Dramatic, with hand outstretched' (Wright).

[3] Scrooge's childhood and schooldays, much amended in *Berg*, are here eventually all deleted: the first big cut.

[4] Marginal stage-direction *Scrooge melted*; also (below) *Melted* both at 'Why, it's old Fezziwig!' and at 'Dick Wilkins, to be sure!' Scrooge's 'melting' had been evident at the end of the scene with his sister, when that was performed: *Soften very much* at 'So she had. You're right ...'

[5] Doubly underlined.

'Prentice! Bless me, yes. There he is. He was very much attached to me, was Dick. Poor Dick! Dear, dear!'

'Yo ho, my boys!' said Fezziwig.[1] 'No more work to-night. Christmas Eve, Dick. Christmas, Ebenezer! Let's have the shutters up, before a man can say, Jack Robinson! Clear away, my lads, and let's have lots of room here!'

Clear away! There was nothing they wouldn't have cleared away, or couldn't have cleared away, with old Fezziwig looking on. It was done in a minute. Every movable was packed off, as if it were dismissed from public life for evermore; the floor was swept and watered, the lamps were trimmed, fuel was heaped upon the fire; and the warehouse was as snug, and warm, and dry, and bright a ball-room, as you would desire to see upon a winter's night.

[2]In came a fiddler with a music-book, and went up to the lofty desk, and made an orchestra of it, and tuned like fifty stomach-aches. In came Mrs. Fezziwig, one vast substantial smile. In came the three Miss Fezziwigs, beaming and lovable. In came the six young followers whose hearts they broke.[3] In came all the young men and women employed in the business. In came the housemaid, with her cousin, the baker. In came the cook, with her brother's particular friend, the milkman. In they all came, one after another; some shyly, some boldly, some gracefully, some awkwardly, some pushing, some pulling; in they all came, anyhow and everyhow. Away they all went, twenty couple at once, hands half round and back again the other way; down the middle and up again; round and round in various stages of affectionate grouping; old top couple always turning up in the wrong place; new top couple starting off again, as soon as they got there; all top couples at last, and not a bottom one to help them. When this result was brought about, old Fezziwig, clapping his hands to stop the dance, cried out, 'Well done!' and the fiddler plunged his hot face into a pot of porter, especially provided for that purpose.[4]

There were more dances, and there were forfeits, and more dances, and there was cake, and there was negus, and there was a great piece of Cold Roast, and there was a great piece of Cold Boiled, and there were mince-pies, and plenty of beer. But the great effect of the evening came after the Roast and Boiled, when the fiddler struck up 'Sir Roger de

[1] 'Laughing' (Wright).

[2] Marginal stage-direction *Cheerful narrative.*

[3] The 'six young followers' were increased to 'ten'; similarly, the number of pairs of dancers was amplified, 'twenty' becoming 'thirty', etc. (Hill).

[4] '. . . a pot of porter, *which positively hissed* . . .' (Hill). As Fezziwig, Dickens clapped hands; at 'Sir Roger de Coverley' (below), he fiddled (Wright). The dance was made vivid through 'the incomparable action of his hands. They actually perform upon the table, as if it were the floor of Fezziwig's room, and every finger were a leg belonging to one of the Fezziwig family' (Field, p. 31).

Coverley.' Then old Fezziwig stood out to dance with Mrs. Fezziwig. Top couple too; with a good stiff piece of work cut out for them; three or four and twenty pair of partners; people who were not to be trifled with; people who *would*[1] dance, and had no notion of walking.

But if they had been twice as many—four times—old Fezziwig would have been a match for them, and so would Mrs. Fezziwig. As to *her*, she was worthy to be his partner in every sense of the term. A positive light appeared to issue from Fezziwig's calves. They shone in every part of the dance. You could n't have predicted, at any given time, what would become of 'em next. And when old Fezziwig and Mrs. Fezziwig had gone all through the dance; advance and retire, turn your partner; bow and curtsey; corkscrew; thread-the-needle, and back again to your place; Fezziwig 'cut'—cut so deftly, that he appeared to wink with his legs.[2]

When the clock struck eleven, this domestic ball broke up. Mr. and Mrs. Fezziwig took their stations, one on either side the door, and shaking hands with every person individually as he or she went out, wished him or her a Merry Christmas. When everybody had retired but the two 'prentices, they did the same to them; and thus the cheerful voices died away, and the lads were left to their beds; which were under a counter in the back-shop. †

'A small matter,' said the Ghost, 'to make these silly folks so full of gratitude. He has spent but a few pounds of your mortal money: three or four, perhaps. Is that so much that he deserves this praise?'

'It is n't that,' said Scrooge, heated by the remark, and speaking unconsciously like his former, not his latter, self. 'It is n't that, Spirit. He has the power to render us happy or unhappy; to make our service light or burdensome; a pleasure or a toil. Say that his power lies in words and looks; in things so slight and insignificant that it is impossible to add and count 'em up: what then? The happiness he gives, is quite as great as if it cost a fortune.'

He felt the Spirit's glance, and stopped.

'What is the matter?'

'Nothing particular.'[3]

'Something, I think?'

'No. No. I should like to be able to say a word or two to my clerk just now! That's all.'

[1] *would*, and *her* (below), italic in *1843*.

[2] A much-praised moment: 'the greatest hit of the evening . . . The contagion of the audience's laughter reached Mr. Dickens himself who with difficulty brought out the inimitable drollery;' his *wink* 'was too much for Boston, and I thought the roof would go off' (*New York Tribune*, 3 December 1867). At the climactic word, 'Mr. Dickens . . . actually did wink with his eyes' (*New York Times*, 10 December 1867).

[3] 'Left finger to mouth' (Wright).

'My time grows short,' observed the Spirit. 'Quick!'

This was not addressed to Scrooge, or to any one whom he could see, but it produced an immediate effect. For again he saw himself. He was older now; a man in the prime of life.

He was not alone, but sat by the side of a fair young girl in a black dress: in whose eyes there were tears.

'It matters little,' she said, softly, to Scrooge's former self. 'To you, very little. Another idol has displaced *me*; and if it can comfort you in time to come, as I would have tried to do, I have no just cause to grieve.'

'What Idol has displaced you?'

'A golden one.[1] I have seen your nobler aspirations fall off one by one, until the master-passion, Gain, engrosses you. Have I not?'

'What then? Even if I have grown so much wiser, what then? I am not changed towards you. † Have I ever sought release from our engagement?'

'In words. No. Never.'

'In what, then?'

'In a changed nature; in an altered spirit; in another atmosphere of life; another Hope as its great end. † If you were free to-day, to-morrow, yesterday, can even I believe that you would choose a dowerless girl: or, choosing her, do I not know that your repentance and regret would surely follow? I do; and I release you. With a full heart, for the love of him you once were.'[2] †

'Spirit! remove me from this place.'

'I told you these were shadows of the things that have been,' said the Ghost. 'That they are what they are, do not blame me!'

'Remove me!' Scrooge exclaimed. 'I cannot bear it! Leave me! Take me back. Haunt me no longer!'[3] †

As he struggled with the Spirit he was conscious of being exhausted, and overcome by an irresistible drowsiness; and, further, of being in his own bedroom. He had barely time to reel to bed, before he sank into a heavy sleep.

[1] *T & F* here includes the sentence, later deleted: 'You fear the world too much'.

[2] The vision of Scrooge's lost love, happy in her marriage to another, is here, after sundry deletions, omitted altogether, both in *Berg* and in *T & F*. But Hill records his performing (and amplifying) part of the omitted episode: 'Spirit, show me no more *of the past*. . . . Why do you delight to torture me? *I cannot bear to see it*" (the latter words in keen agony)—and he went on—"Remove me . . ."'

[3] The paragraph following in *1843* ('In the struggle') is deleted in pencil in *Berg*; there is a marginal *Stet*, in pencil, but the paragraph is also deleted in blue ink. *T & F* does not include the paragraph. The blue-ink deletion is later, and more authoritative than, the uncancelled *Stet*.

STAVE THREE[1]

The Second of the Three Spirits

SCROOGE awoke in his own Bedroom. There was no doubt about that. But *it*, and his own adjoining sitting-room into which he shuffled in his slippers—attracted by a great Light there—had undergone a surprising transformation. The walls and ceiling were so hung with living green, that it looked a perfect grove. The leaves of holly, mistletoe, and ivy reflected back the light, as if so many little mirrors had been scattered there; and such a mighty blaze went roaring up the chimney, as that petrifaction of a hearth had never known in Scrooge's time, or Marley's, or for many and many a winter season gone. Heaped upon the floor, to form a kind of throne, were turkeys, geese, game, brawn, great joints of meat, sucking-pigs, long wreaths of sausages, mince-pies, plum-puddings, barrels of oysters, red-hot chesnuts, cherry-cheeked apples, juicy oranges, luscious pears, immense twelfth-cakes, and great bowls of punch. In easy state upon this couch, there sat a Giant, glorious to see; who bore a glowing torch, in shape not unlike Plenty's horn, and who raised it high, to shed its light on Scrooge, as he came peeping round the door.

'Come in! Come in! and know me better, man! I am the Ghost of Christmas Present. Look upon me! † You have never seen the like of me before!'

'Never.'[2]

'Have never walked forth with the *younger* members of my family; meaning (for I am very young) my elder brothers born in these later years?' pursued the Phantom.

'I don't think I have. I am afraid I have not. Have you had many brothers, Spirit?'

'More than eighteen hundred.'

'A tremendous family to provide for! Spirit, conduct me where you will. I went forth last night on compulsion, and I learnt a lesson which is working now. To-night, if you have aught to teach me, let me profit by it.'

'Touch my robe!'

Scrooge did as he was told, and held it fast.

The room, and its contents, all vanished instantly, and they stood in the city streets upon a snowy Christmas morning.[3] †

[1] '*Chapter 3*' in *Berg*, regularized in *T & F* to STAVE THREE. After tinkering with its opening, Dickens deleted the first three pages.

[2] 'Scrooge did *not* say "Never"; but (in a trembling voice),—"Well, I don't think I have"' (Hill).

[3] Another lengthy deletion here—the lively description of the streets and shops, and of people going to church. By contrast, the Cratchit episode which follows is hardly abbreviated at all.

Scrooge and the Ghost passed on, invisible,[1] straight to Scrooge's clerk's; and on the threshold of the door the Spirit smiled, and stopped to bless Bob Cratchit's dwelling with the sprinklings of his torch. Think of that! Bob had but fifteen 'Bob' a-week himself; he pocketed on Saturdays but fifteen copies of his Christian name; and yet the Ghost of Christmas Present blessed his four-roomed house!

Then up rose Mrs. Cratchit, Cratchit's wife, dressed out but poorly in a twice-turned gown, but brave in ribbons—which are cheap and make a goodly show for sixpence; and she laid the cloth, assisted by Belinda Cratchit, second of her daughters, also brave in ribbons; while Master Peter Cratchit plunged a fork into the saucepan of potatoes, and getting the corners of his monstrous shirt-collar (Bob's private property, conferred upon his son and heir in honour of the day) into his mouth, rejoiced to find himself so gallantly attired, and yearned to show his linen in the fashionable Parks. And now two smaller Cratchits, boy and girl, came tearing in, screaming that, outside the baker's, they had smelt the goose, and known it for their own; and basking in luxurious thoughts of sage-and-onion, these young Cratchits danced about the table, and exalted Master Peter Cratchit to the skies, while he (not proud, although his collars nearly choked him) blew the fire, until the slow potatoes bubbling up, knocked loudly at the saucepan-lid to be let out and peeled.

'What has ever got your precious father, then?' said Mrs. Cratchit. 'And your brother, Tiny Tim! And Martha warn't as late last Christmas Day by half-an-hour!'

'Here's Martha, mother!' said a girl, appearing as she spoke.

'Here's Martha, mother!' cried the two young Cratchits. 'Hurrah! There's *such*[2] a goose, Martha!'

'Why, bless your heart alive, my dear, how late you are!' said Mrs. Cratchit, kissing her a dozen times, and taking off her shawl and bonnet for her.

'We'd a deal of work to finish up last night,' replied the girl, 'and had to clear away this morning, mother!'

'Well! Never mind so long as you are come,' said Mrs. Cratchit. 'Sit ye down before the fire, my dear, and have a warm, Lord bless ye!'

'No no! There's father coming,' cried the two young Cratchits, who were everywhere at once. 'Hide, Martha, hide!'

So Martha hid herself, and in came little Bob, the father,[3] with at least three feet of comforter exclusive of the fringe, hanging down before him; and his thread-bare clothes darned up and brushed, to look seasonable;

[1] Doubly underlined (probably because these words, written in the margin, marked the resumption of text after a long cut).

[2] *such* italic in *1843*. 'The way those two young Cratchits hail Martha, and exclaim [this] . . . can never be forgotten' (Field, p. 32).

[3] Marginal stage-direction *Tone to Tiny Tim.*

and Tiny Tim upon his shoulder. *Alas for Tiny Tim, he bore a little crutch, and had his limbs supported by an iron frame*![1]

'Why, where's our Martha?' cried Bob Cratchit, looking round.

'Not coming,' said Mrs. Cratchit.

'Not coming!' said Bob, with a sudden declension in his high spirits; for he had been Tim's blood horse all the way from church, and had come home rampant. 'Not coming upon Christmas Day!'

Martha didn't like to see him disappointed, if it were only in joke; so she came out prematurely from behind the closet door, and ran into his arms, while the two young Cratchits hustled Tiny Tim, and bore him off into the wash-house, that he might hear the pudding singing in the copper.

'And how did little Tim behave?' asked Mrs. Cratchit, when she had rallied Bob on his credulity and Bob had hugged his daughter to his heart's content.

'As good as gold,' said Bob, 'and better. Somehow he gets thoughtful, sitting by himself so much, and thinks the strangest things you ever heard. He told me, coming home, that he hoped the people saw him in the church, because he was a cripple, and it might be pleasant to them to remember upon Christmas Day, who made lame beggars walk and blind men see.'[2]

Bob's voice was tremulous when he told them this, and trembled more when he said that Tiny Tim was growing strong and hearty.

His active little crutch was heard upon the floor, and back came Tiny Tim before another word was spoken, escorted by his brother and sister to his stool beside the fire; and while Bob, turning up his cuffs—as if, poor fellow, they were capable of being made more shabby—compounded some hot mixture in a jug with gin and lemons, and stirred it round and round and put it on the hob to simmer; Master Peter and the two ubiquitous young Cratchits went to fetch the goose, with which they soon returned in high procession.

Mrs. Cratchit made the gravy (ready beforehand in a little saucepan) hissing hot;[3] Master Peter mashed the potatoes with incredible vigour; Miss Belinda sweetened up the apple-sauce; Martha dusted the hot plates; Bob took Tiny Tim beside him in a tiny corner at the table; the two young Cratchits set chairs for everybody, not forgetting themselves, and mounting guard upon their posts, crammed spoons into their mouths, lest they should shriek for goose before their turn came to be helped. At last the dishes were set on, and grace was said. It was succeeded by a

[1] Doubly underlined.

[2] This speech 'Feelingly, tremblingly' and 'With hkf' at 'blind men see' (Wright); its pathos was 'the most delicate and artistic rendering of the whole reading' (Field, p. 32).

[3] 'Mr Dickens . . . is one of the best of living actors,' wrote the *New York Times* critic (10 December 1867): and, instancing his 'free use of gesticulation', he described how, in this passage, he stirred the gravy, mashed the potatoes, dusted the plates, and (later) sniffed the famous pudding.

breathless pause, as Mrs. Cratchit, looking slowly all along the carving-knife, prepared to plunge it in the breast; but when she did, and when the long expected gush of stuffing issued forth, one murmur of delight arose all round the board, and even Tiny Tim, excited by the two young Cratchits, beat on the table with the handle of his knife, and feebly cried Hurrah![1]

There never was such a goose. Bob said he didn't believe there ever was such a goose cooked. Its tenderness and flavour, size and cheapness, were the themes of universal admiration. Eked out by the apple-sauce and mashed potatoes, it was a sufficient dinner for the whole family; indeed, as Mrs. Cratchit said with great delight (surveying one small atom of a bone upon the dish), they hadn't ate it all at last! Yet every one had had enough, and the youngest Cratchits in particular, were steeped in sage and onion to the eyebrows! But now, the plates being changed by Miss Belinda, Mrs. Cratchit left the room alone—too nervous to bear witnesses—to take the pudding up, and bring it in.

Suppose it should not be done enough![2] Suppose it should break in turning out! Suppose somebody should have got over the wall of the back-yard, and stolen it, while they were merry with the goose: a supposition at which the two young Cratchits became livid! All sorts of horrors were supposed.

Hallo! A great deal of steam! The pudding was out of the copper. A smell like a washing-day! That was the cloth. A smell like an eating-house, and a pastry cook's next door to each other, with a laundress's next door to that! That was the pudding! In half a minute Mrs. Cratchit entered: flushed, but smiling proudly: with the pudding, like a speckled cannon-ball, so hard and firm, blazing in half of half-a-quartern of ignited brandy, and bedight with Christmas holly stuck into the top.

Oh, a wonderful pudding![3] Bob Cratchit said, and calmly too, that he regarded it as the greatest success achieved by Mrs. Cratchit since their marriage. Mrs. Cratchit said that now the weight was off her mind, she would confess she had had her doubts about the quantity of flour. Everybody had something to say about it, but nobody said or thought it was at all a small pudding for a large family. Any Cratchit would have blushed to hint at such a thing.

At last the dinner was all done, the cloth was cleared, the hearth swept, and the fire made up. The compound in the jug being tasted and considered perfect, apples and oranges were put upon the table, and a

[1] 'Very high thin voice' (Wright).

[2] 'Shaking to and fro with his body' during this paragraph (Wright).

[3] 'What cheers when Mrs. Cratchit brought in that pudding. . . . His description brought down torrents of applause, so archly was it given' (*Cambridge Independent Press*, 17 October 1859). Dickens's sniffing the pudding made Kate Field remark (p. 33) that 'What Dickens *does* is frequently infinitely better than anything he says, or the way he says it.'

shovel-full of chesnuts on the fire.[1] Then all the Cratchit family drew round the hearth, in what Bob Cratchit called a circle; and at Bob Cratchit's elbow stood the family display of glass; two tumblers, and a custard-cup without a handle.[2]

These held the hot stuff from the jug, however, as well as golden goblets would have done; and Bob served it out with beaming looks, while the chesnuts on the fire sputtered and crackled noisily. Then Bob proposed:

'A Merry Christmas to us all, my dears. God bless us!'

Which all the family re-echoed.

'God bless us every one!' said Tiny Tim, the last of all.

He sat very close to his father's side, upon his little stool. Bob held his withered little hand in his, as if he loved the child, and wished to keep him by his side, and dreaded that he might be taken from him.[3] †

Scrooge raised his head speedily, on hearing his own name.

'Mr. Scrooge!' said Bob; 'I'll give you Mr. Scrooge, the Founder of the Feast!'

'The Founder of the Feast indeed!' cried Mrs. Cratchit, reddening. 'I wish I had him here. I'd give him a piece of my mind to feast upon, and I hope he'd have a good appetite for it.'

'My dear,' said Bob, 'the children! Christmas Day.'

'It should be Christmas Day, I am sure,' said she, 'on which one drinks the health of such an odious, stingy, hard, unfeeling man as Mr. Scrooge. You know he is, Robert! Nobody knows it better than you do, poor fellow!'

'My dear,' was Bob's mild answer, 'Christmas Day.'

'I'll drink his health for your sake and the Day's,' said Mrs. Cratchit, 'not for his. Long life to him! A merry Christmas and a happy new year! He'll be very merry and very happy, I have no doubt!'[4]

The children drank the toast after her. It was the first of their proceedings which had no heartiness in it. Tiny Tim drank it last of all, but he didn't care twopence for it. Scrooge was the Ogre of the family. The mention of his name cast a dark shadow on the party, which was not dispelled for full five minutes.

After it had passed away, they were ten times merrier than before, from the mere relief of Scrooge the Baleful being done with. Bob Cratchit told them how he had a situation in his eye for Master Peter, which would

[1] *T & F* begins a new paragraph here.

[2] '. . . two *broken* tumblers, and a *cracked* custard cup . . .' (Hill).

[3] 'Who can forget Bob Cratchit, holding Tiny Tim's hand, then throwing him a kiss, and brushing a tear from his eyes, as he prepares to propose the health of Scrooge? It was a little action, but it meant so much! Those only who have children and fear to lose them, or loving them *have* lost, can know how much it meant' (*New York Tribune*, 10 December 1867). Cf. Field, p. 34, on this passage.

[4] She drank the toast 'sharply' (Hill).

bring in, if obtained, full five-and-sixpence weekly. The two young Cratchits laughed tremendously at the idea of Peter's being a man of business; and Peter himself looked thoughtfully at the fire from between his collars, as if he were deliberating what particular investments he should favour when he came into the receipt of that bewildering income. Martha, who was a poor apprentice at a milliner's, then told them what kind of work she had to do, and how many hours she worked at a stretch, and how she meant to lie a-bed to-morrow morning for a good long rest; to-morrow being a holiday she passed at home. Also how she had seen a countess and a lord some days before, and how the lord 'was much about as tall as Peter;' at which Peter pulled up his collars so high that you couldn't have seen his head if you had been there. All this time the chesnuts and the jug went round and round; and bye and bye they had a song, about a lost child travelling in the snow, from Tiny Tim; who had a plaintive little voice, and sang it very well indeed.

There was nothing of high mark in this. They were not a handsome family; they were not well dressed; their shoes were far from being waterproof; their clothes were scanty; and Peter might have known, and very likely did, the inside of a pawnbroker's. But they were happy, grateful, pleased with one another, and contented with the time; and when they faded, and looked happier yet in the bright sprinklings of the Spirit's torch at parting, Scrooge had his eye upon them, and especially on Tiny Tim, until the last.[1] †

It was a great surprise to Scrooge, *as this scene vanished*,[2] to hear a hearty laugh. It was a much greater surprise to Scrooge to recognize it as his own nephew's, and to find himself in a bright, dry, gleaming room, with the Spirit standing smiling by his side, and looking at that same nephew.

It is a fair, even-handed, noble adjustment of things, that while there is infection in disease and sorrow, there is nothing in the world so irresistibly contagious as laughter and good-humour. When Scrooge's nephew laughed, Scrooge's niece, by marriage, laughed as heartily as he. And their assembled friends being not a bit behindhand, laughed out, lustily.

'He said that Christmas was a humbug, as I live!' cried Scrooge's nephew. 'He believed it too!'[3]

'More shame for him, Fred!' said Scrooge's niece, indignantly.—

[1] Scrooge's vision of the miners, lighthouse-keepers, and mariners is here deleted.

[2] A manuscript insertion, doubly underlined (cf. p. 19, note 1, above). Marginal stage-direction here, *Tone to cheerful Narrative*.

[3] Scrooge's nephew's laughter here was 'contagious' (Hill), and the ensuing account of his niece was 'irresistibly exhilarating' (Kent, p. 104). 'I hate him!' she added, after the text's 'More shame for him, Fred!' and Fred, instead of calling Scrooge 'a comical old fellow' called him 'a queer old fellow'—a great improvement (Hill noted and remarked).

Bless those women; they never do anything by halves. They are always in earnest.

She was very pretty: exceedingly pretty. With a dimpled, surprised-looking, capital face; a ripe little mouth, that seemed made to be kissed—as no doubt it was; all kinds of good little dots about her chin, that melted into one another when she laughed; and the sunniest pair of eyes you ever saw in any little creature's head. Altogether she was what you would have called provoking; but satisfactory, too. Oh, perfectly satisfactory!

'He's a comical old fellow,' said Scrooge's nephew, 'that's the truth; and not so pleasant as he might be. However, his offences carry their own punishment, and I have nothing to say against him. † Who suffers by his ill whims? Himself, always. Here, he takes it into his head to dislike us, and he won't come and dine with us. What's the consequence? He don't lose much of a dinner.'

'Indeed, I think he loses a very good dinner,' interrupted Scrooge's niece. Everybody else said the same, and they must be allowed to have been competent judges, because they had just had dinner; and, with the dessert upon the table, were clustered round the fire, by lamplight.

'Well! I am very glad to hear it,' said Scrooge's nephew, 'because I haven't any great faith in these young housekeepers. What do *you*[1] say, Topper?'

Topper clearly had his eye on one of Scrooge's niece's sisters, for he answered that a bachelor was a wretched outcast, who had no right to express an opinion on the subject. Whereat Scrooge's niece's sister—the plump one with the lace tucker: not the one with the roses—blushed. †

After tea, they had some music. For they were a musical family, and knew what they were about, when they sung a Glee or Catch, I can assure you: especially Topper, who could growl away in the bass like a good one, and never swell the large veins in his forehead, or get red in the face over it.[2] † But they didn't devote the whole evening to music. After a while they played at forfeits; *for it is good to be children sometimes, and never better than at Christmas, when its mighty Founder was a child himself.* There was first a game at blind-man's buff though. And I no more believe Topper was really blinded than I believe he had eyes in his boots.[3] *Because, the way in which*[4] he went after that plump sister in the lace tucker, was an

[1] *you* italic in *1843*.

[2] Against the deleted passage which follows, in which Scrooge 'softened more and more', marginal stage-direction *Tone down to Pathos*; but *Up to cheerfulness* at 'But they didn't devote the whole evening to music'.

[3] An episode 'never to be forgotten'. When Dickens spoke this sentence, 'his facial expression—indignant as of a man who is being put upon, and yet with a consciousness of the absurdity of the statement that makes him laugh in spite of his anger—was inimitable, and it was long before the audience would let him get on.' The plump sister was 'immortal' (*New York Tribune*, 3 December 1867).

[4] A manuscript insertion, doubly underlined (cf. p. 19, note 1).

outrage on the credulity of human nature. Knocking down the fire-irons, tumbling over the chairs, bumping up against the piano, smothering himself among the curtains, wherever she went, there went he. He always knew where the plump sister was. He wouldn't catch anybody else. If you had fallen up against him, as some of them did, and stood there; he would have made a feint of endeavouring to seize you, which would have been an affront to your understanding; and would instantly have sidled off in the direction of the plump sister. †

'Here is a new game,' said Scrooge.[1] 'One half hour, Spirit, only one!'

It was a Game called Yes and No, where Scrooge's nephew had to think of something, and the rest must find out what; he only answering to their questions yes or no as the case was. The fire of questioning to which he was exposed, elicited from him that he was thinking of an animal, a live animal, rather a disagreeable animal, a savage animal, an animal that growled and grunted sometimes, and talked sometimes, and lived in London, and walked about the streets, and wasn't made a show of, and wasn't led by anybody, and didn't live in a menagerie, and was never killed in a market, and was not a horse, or an ass, or a cow, or a bull, or a tiger, or a dog, or a pig, or a cat, or a bear. At every new question put to him, this nephew burst into a fresh roar of laughter; and was so inexpressibly tickled, that he was obliged to get up off the sofa and stamp. At last the plump sister cried out:

'I have found it out! I know what it is, Fred! I know what it is!'

'What is it?' cried Fred.

'It's your uncle Scro-o-o-o-oge!'[2]

Which it certainly was. Admiration was the universal sentiment, though some objected that the reply to 'Is it a bear?' ought to have been 'Yes.' †

Uncle Scrooge had imperceptibly become so gay and light of heart, that he would have drank to the unconscious company in an inaudible speech. But the whole scene passed off in the breath of the last word spoken by his nephew; and he and the Spirit were again upon their travels.

Much they saw, and far they went, and many homes they visited, but always with a happy end. The Spirit stood beside sick beds, and they were cheerful; on foreign lands, and they were close at home; by struggling men, and they were patient in their greater hope; by poverty, and it was rich. In almshouse, hospital, and jail, in misery's every refuge, where vain man in his little brief authority had not made fast the door, and barred

[1] 'This scene he treated quite freely,' notes Hill, who records Dickens's textual variations hereabouts. This speech 'Laughingly and cryingly' (Wright).

[2] 'In high key' (Wright); an 'abiding memory' for many witnesses—'the blood-curdling and yet almost loving way in which the name . . . was pronounced—long-drawn-out and with tremendous emphasis' (Walter Pine, in *Dickensian*, xxxiv (1939), 206)

the Spirit out, he left his blessing, and taught Scrooge his precepts.[1] Suddenly, as they stood together in an open place, the bell struck *Twelve.*[2] †

Scrooge looked about him for the Ghost, and saw it no more. As the last stroke ceased to vibrate, he remembered the prediction of old Jacob Marley, and lifting up his eyes, beheld a solemn Phantom, draped and hooded, coming, like a mist along the ground, towards him.

STAVE FOUR

The Last of the Spirits

THE Phantom slowly, gravely, silently, approached.[3] When it came near him, Scrooge bent down upon his knee; for in the air through which this Spirit moved it seemed to scatter gloom and mystery.

It was shrouded in a deep black garment, which concealed its head, its face, its form, and left nothing of it visible *save one outstretched hand.*[4] He knew no more, for the Spirit neither spoke nor moved.

'I am in the presence of the Ghost of Christmas Yet To Come? † Ghost of the Future! I fear you more than any Spectre I have seen. But, as I know your purpose is to do me good, and as I hope to live to be another man from what I was, I am prepared to bear you company, and do it with a thankful heart.[5] Will you not speak to me?'

It gave him no reply. The hand was pointed straight before them.

'Lead on! Lead on! The night is waning fast, and it is precious time to me, I know. Lead on, Spirit!'

They scarcely seemed to enter the city; for the city rather seemed to spring up about them. But there they were, in the heart of it; on Change, amongst the merchants.

The Spirit stopped beside one little knot of business men. Observing that the hand was pointed to them, Scrooge advanced to listen to their talk.

'No,' said a great fat man with a monstrous chin, 'I don't know much about it, either way. I only know he's dead.'[6]

[1] 'Pause. Look from one side to the other' (Wright).

[2] The encounter with the two terrible children, Ignorance and Want (preluded by the marginal stage-direction *Stern Pathos*), is here deleted.

[3] Marginal stage-direction *Mystery*: and, above STAVE FOUR, *Throughout*, *Monotonous Hand.*

[4] Here 'he stretched out his right arm at full length, and soon bent the hand down very definitely' (Hill).

[5] 'A very long pause' here: 'Dickens's face looked so anxious that the audience watched him in breathless silence. "Will you not *speak* to me?" with great emphasis on the one word "speak"' (Hill).

[6] 'Low guttural voice' (Wright). This fat man was often mentioned as an instance of Dickens's power of giving an unforgettable individuality to minor characters; e.g., see Kent, p. 96.

'When did he die?' inquired another.

'Last night, I believe.'

'Why, what was the matter with him? I thought he'd never die.'

'God knows,' said the first, with a yawn.

'What has he done with his money?' asked a red-faced gentleman.

'I haven't heard,' said the man with the large chin. 'Company, perhaps. He hasn't left it to *me*.[1] That's all I know. Bye, bye!' †

Scrooge was at first inclined to be surprised that the Spirit should attach importance to conversation apparently so trivial; but feeling assured that it must have some hidden purpose, he set himself to consider what it was likely to be. It could scarcely be supposed to have any bearing on the death of Jacob, his old partner, for that was Past, and this Ghost's province was the Future.

He looked about in that very place for his own image; *but another man stood in his accustomed corner*,[2] and though the clock pointed to his usual time of day for being there, *he saw no likeness of himself* among the multitudes that poured in through the Porch. It gave him little surprise, however; for he had been revolving in his mind a change of life, and he thought and hoped he saw his new-born resolutions carried out in this.

They left this busy scene, and went into an obscure part of the town, to a low shop where iron, old rags, bottles, bones, and greasy offal, were bought by a gray-haired rascal, of great age; who sat smoking his pipe.

Scrooge and the Phantom came into the presence of this man, just as a woman with a heavy bundle slunk into the shop. But she had scarcely entered, when another woman, similarly laden, came in too; and she was closely followed by a man in faded black. After a short period of blank astonishment, in which the old man with the pipe had joined them, they all three burst into a laugh.[3]

'Let the charwoman alone to be the first!' cried she who had entered first. 'Let the laundress alone to be the second; and let the undertaker's man alone to be the third. Look here, old Joe, here's a chance! If we haven't all three met here without meaning it!'

'You couldn't have met in a better place. You were made free of it long ago, you know; and the other two an't strangers. What have you got to sell, what have you got to sell?'

'Half a minute's patience, Joe, and you shall see. What odds then! What odds, Mrs. Dilber?' said the woman. 'Every person has a right to take care of themselves. *He*[4] always did! Who's the worse for the loss of a few things like these? Not a dead man, I suppose.'

[1] *me* italic in *1843*.

[2] This, and the next italicized phrase, doubly underlined.

[3] Marginal stage-direction *Weird*. 'There was something positively and Shakespearianly weird in the laugh and tone of the charwoman' (Field, p. 35). 'Miserable high wheezing voice' for Old Joe (Wright).

[4] *He* italic in *1843*.

Mrs. Dilber, whose manner was remarkable for general propitiation, said, 'No, indeed, Ma'am.'

'If he wanted to keep 'em after he was dead, a wicked old screw, why wasn't he natural in his lifetime? If he had been, he'd have had somebody to look after him when he was struck with Death, instead of lying gasping out his last there, alone by himself.'

'It's the truest word that ever was spoke. It's a judgement on him.'

'I wish it was a little heavier judgement; and it should have been, you may depend upon it, if I could have laid my hands on anything else. Open that bundle, old Joe, and let me know the value of it. Speak out plain. I'm not afraid to be the first, nor afraid for them to see it.' †

Joe went down on his knees for the greater convenience of opening the bundle, and dragged out a large and heavy roll of some dark stuff.

'What do you call this? Bed-curtains!'

'Ah! Bed-curtains! Don't drop that oil upon the blankets, now.'

'*His* blankets?'

'Whose else's do you think? He isn't likely to take cold without 'em, I dare say. Ah! You may look through that shirt till your eyes ache; but you won't find a hole in it, nor a threadbare place. It's the best he had, and a fine one too. They'd have wasted it by dressing him up in it, if it hadn't been for me.'

Scrooge listened to this dialogue in horror.

'Spirit! I see, I see. The case of this unhappy man might be my own. My life tends that way, now. *Merciful Heaven, what is this!*'[1]

The scene had changed, and now he almost touched a bare, uncurtained bed. A pale light, rising in the outer air, fell straight upon this bed; and on it, unwatched, unwept, uncared for, was the body of this plundered man unknown.[2] †

'*Spirit! Let me see some tenderness connected with a death*, or this dark chamber, Spirit, will be for ever present to me.'

[3]The Ghost conducted him to poor Bob Cratchit's house; the dwelling he had visited before; and found the mother and the children seated round the fire.

Quiet. Very quiet. The noisy little Cratchits were as still as statues in one corner, and sat looking up at Peter, who had a book before him. The mother and her daughters were engaged in needlework. But surely they were very quiet!

'"And He took a child, and set him in the midst of them."'

Where had Scrooge heard those words? He had not dreamed them.

[1] Doubly underlined. Marginal stage-direction *Scrooge's start and change to terror.*

[2] 'Pause. Look to left' (Wright). *T & F* has '... plundered unknown man'. In the next line '*Spirit*!' (a marginal insertion) is doubly underlined, the rest singly.

[3] Marginal stage-direction *Pathos*. 'Looking about' (Wright).

The boy must have read them out, as he and the Spirit crossed the threshold. Why did he not go on?

The mother laid her work upon the table, and put her hand up to her face.

'The colour hurts my eyes,' she said.[1]

The colour? *Ah, poor Tiny Tim!*

'They're better now again. It makes them weak by candle-light; and I wouldn't show weak eyes to your father when he comes home, for the world. It must be near his time.'

'Past it rather,' Peter answered, shutting up his book. 'But I think he has walked a little slower than he used, these few last evenings, mother.'

'I have known him walk with—I have known him walk with Tiny Tim upon his shoulder, very fast indeed.'

'And so have I,' cried Peter. 'Often.'

'And so have I,' exclaimed another. So had all.

'But he was very light to carry, and his father loved him so, that it was no trouble: no trouble. And there is your father at the door!'

She hurried out to meet him; and little Bob in his comforter—he had need of it, poor fellow—came in. His tea was ready for him on the hob, and they all tried who should help him to it most. Then the two young Cratchits got upon his knees and laid, each child a little cheek, against his face, as if they said, 'Don't mind it, father. Don't be grieved!'

Bob was very cheerful with them, and spoke pleasantly to all the family. He looked at the work upon the table, and praised the industry and speed of Mrs. Cratchit and the girls. They would be done long before Sunday, he said.

'Sunday! You went to-day, then, Robert?'

'Yes, my dear,' returned Bob. 'I wish you could have gone. It would have done you good to see how green a place it is. But you'll see it often. I promised him that I would walk there on a Sunday.[2] My little, little child! My little child!'

He broke down all at once. He couldn't help it. *If he could have helped it, he and his child would have been farther apart perhaps than they were.* †

[1] 'Sigh. Handkerchief to eyes', a gesture repeated below at 'Sunday! . . .' (Wright). 'The pathos which he throws into [this] one short line . . . cannot be described' (*Town Talk*, 1858, quoted in *Dickensian*, xxxvii (1941), 223); a passage 'breathing an exquisite tenderness . . . that thrilled to the hearts of all who heard [it], and still, we doubt not, haunts their recollections' (Kent, p. 104).

[2] '3 or 4 seconds of painful silence. Then Bob's grief burst out, almost in a suppressed scream' (Hill); 'Cry with head thrown back' (Wright). Some critics thought Dickens over-played this moment (e.g., Field, p. 35). He made 'more of a "point" of it than of any other passage' in the Reading (New York *Nation*, 12 December 1867, p. 483). Kent, who praises Dickens's rendering, notes how much of this pathetic episode is deleted, e.g. Bob's visit to the deathbed (pp. 104–5).

'Spectre,' said Scrooge, 'something informs me that our parting moment is at hand. I know it, but I know not how. Tell me what man that was with the covered face whom we saw lying dead?'

The Ghost of Christmas Yet To Come conveyed him † to a dismal, wretched, ruinous churchyard.

The Spirit stood among the graves, and pointed down to One.

'Before I draw nearer to that stone to which you point, answer me one question. *Are these the shadows of the things that Will be, or are they shadows of the things that May be, only?*'[1]

Still the Ghost pointed downward to the grave by which it stood.

'Men's courses will foreshadow certain ends, to which, if persevered in, they must lead. But if the courses be departed from, the ends will change. Say it is thus with what you show me!'

The Spirit was immovable as ever. Scrooge crept towards it, trembling as he went; and following the finger, read upon the stone of the neglected grave his own name, EBENEZER SCROOGE.[2]

'Am *I*[3] that man who lay upon the bed? No, Spirit! Oh no, no! Spirit! hear me! I am not the man I was. I will not be the man I must have been but for this intercourse. Why show me this, if I am past all hope? Assure me that I yet may change these shadows you have shown me, by an altered life.'

For the first time, the kind hand faltered.

'I will honour Christmas in my heart, and try to keep it all the year. I will live in the Past, the Present, and the Future. The Spirits of all Three shall strive within me. I will not shut out the lessons that they teach. Oh, tell me I may sponge away the writing on this stone!'

Holding up his hands in one last prayer to have his fate reversed, he saw an alteration in the Phantom's hood and dress. It shrunk, collapsed, and dwindled down into a bedpost.[4]

Yes! and the bedpost was his own. The bed was his own, the room was his own. Best and happiest of all, the time before him was his own, to make amends in! †

He was checked in his transports by the churches ringing out the lustiest peals he had ever heard. Running to the window, he opened it, and put out

[1] *Will* underlined three times, *May* four times, the rest singly. Dickens also much emphasized the concluding word, 'only' (Hill).

[2] The Ghost's right hand pointed downwards to the grave. 'Dickens put great terror into the tones of Scrooge' in 'Am *I* that man . . .' The silent Ghost's finger continued to point downwards, and in the rest of his speech 'Scrooge's voice lost all its stern, hard character' (Hill). He put both hands up, then clasped them (Wright). Then, at 'I will honour Christmas . . .' his voice 'became fully charged with firm tones and a humaner outlook' (Hill).

[3] *I* italic in *1843*.

[4] The text here runs straight into the story's STAVE FIVE, without any break, Dickens reading its opening paragraph 'in his richest, fullest, happiest voice' (Hill).

his head. No fog, no mist, no night; clear, bright, stirring, golden Day!

'What's to-day?' cried Scrooge, calling downward to a boy in Sunday clothes, who perhaps had loitered in to look about him.

'EH?'[1]

'What's to-day, my fine fellow?'

'To-day! Why, CHRISTMAS DAY.'

'It's Christmas Day! I haven't missed it. Hallo, my fine fellow!'

'Hallo!'

'Do you know the Poulterer's, in the next street but one, at the corner?'

'I should hope I did.'

'An intelligent boy! A remarkable boy! Do you know whether they've sold the prize Turkey that was hanging up there? Not the little prize Turkey; the big one?'

'What, the one as big as me?'

'What a delightful boy! It's a pleasure to talk to him. Yes, my buck!'

'It's hanging there now.'

'Is it? Go and buy it.'

'Walk-ER!'[2] exclaimed the boy.

'No, no, I am in earnest. Go and buy it, and tell 'em to bring it here, that I may give them the direction where to take it. Come back with the man, and I'll give you a shilling. Come back with him in less than five minutes, and I'll give you half-a-crown!'

The boy was off like a shot.

'I'll send it to Bob Cratchit's! He sha'n't know who sends it. It's the size of Tiny Tim. Joe Miller never made such a joke as sending it to Bob's will be!'

The hand in which he wrote the address was not a steady one, but write it he did, somehow, and went down stairs to open the street door, ready for the coming of the poulterer's man.

It *was*[3] a Turkey! He never could have stood upon his legs, that bird. He would have snapped 'em short off in a minute, like sticks of sealing-wax. †

He[4] dressed himself 'all in his best,' and at last got out into the streets. The people were by this time pouring forth, as he had seen them with

[1] Capitals here, and in the boy's next speech, in *1843*: also in his 'Walk-ER' below. 'Boy's voice very high' (Wright). A famous moment: every word of it was 'watched for and listened to by audiences like celebrated passages from a great standard play' (*Critic*, 4 September 1858, p. 537). After one performance, Dickens wrote to Forster: 'if you could have seen [the audience during this dialogue] . . . I doubt if you would ever have forgotten it' (*N*, iii. 62). He sometimes amplified the dialogue, e.g., 'What a conversational boy!' (Hill).

[2] 'Dickens, as the boy, put his thumb to his nose, and spread out his fingers, with a jeer' (Hill).

[3] *was* italic in *1843*.

[4] 'Scrooge' in *T & F*: presumably Dickens's correction.

the Ghost of Christmas Present; and walking with his hands behind him, Scrooge regarded every one with a delighted smile. He looked so irresistibly pleasant, in a word, that three or four good-humoured fellows said, 'Good morning, sir! A merry Christmas to you!' And Scrooge said often afterwards, that of all the blithe sounds he had ever heard, those were the blithest in his ears. †

In the afternoon, he turned his steps towards his nephew's house.

He passed the door a dozen times, before he had the courage to go up and knock. But he made a dash, and did it.

'Is your master at home, my dear?' said Scrooge to the girl. Nice girl! Very.[1]

'Yes, sir.'

'Where is he, my love?'

'He's in the dining-room, sir, along with mistress.'

'He knows me,' said Scrooge, with his hand already on the dining-room lock. 'I'll go in here, my dear. Fred!'

'Why bless my soul!' cried Fred, 'who's that?'

'It's I. Your uncle Scrooge. I have come to dinner. Will you let me in, Fred?'

Let him in! It is a mercy he didn't shake his arm off. He was at home in five minutes. Nothing could be heartier. His niece looked just the same. So did Topper when *he*[2] came. So did the plump sister, when *she* came. So did every one when *they* came. Wonderful party, wonderful games, wonderful unanimity, won-der-ful happiness!

But he was early at the office next morning. Oh he was early there. If he could only be there first, and catch Bob Cratchit coming late! That was the thing he had set his heart upon.

And he did it! The clock struck nine. No Bob. A quarter past. No Bob. Bob was full eighteen minutes and a half, behind his time.

Bob's hat was off, before he opened the door; his comforter too. He was on his stool in a jiffy; driving away with his pen, as if he were trying to overtake nine o'clock.

'Hallo!' growled Scrooge, in his accustomed voice as near as he could feign it.[3] 'What do you mean by coming here at this time of day?'

'I am very sorry, sir. I *am*[4] behind my time.'

'You are? Yes. I think you are. Step this way, if you please.'

'It's only once a year, sir. It shall not be repeated. I was making rather merry yesterday, sir.'

[1] 'There was a sort of parenthetical smack of the lips in [this] self-communing of Scrooge . . . "*Nice girl! very!*" Then, as to the cordiality of his reception by his Nephew, what could possibly have expressed it better than the look, voice, manner of the Reader?' (Kent, pp. 106–7).

[2] *he . . . she . . . they* italic in *1843*.

[3] Several reports noted Scrooge's comic inability to 'feign' his old gruff voice.

[4] *am* italic in *1843*. Bob's speech 'full of hesitancy and timidity' (Hill).

'Now, I'll tell you what, my friend. I am not going to stand this sort of thing any longer.—And therefore,' Scrooge continued, leaping from his stool, and giving Bob such a dig in the waistcoat that he staggered back into the Tank again: 'and therefore I am about to raise your salary!'

Bob trembled, and got a little nearer to the ruler.

'A merry Christmas, Bob!' said Scrooge, with an earnestness that could not be mistaken, as he clapped him on the back. 'A merrier Christmas, Bob, my good fellow, than I have given you, for many a year! I'll raise your salary, and endeavour to assist your struggling family, and we will discuss your affairs this very afternoon, over a Christmas bowl of smoking bishop, Bob![1] Make up the fires, and buy a second coal-scuttle before you dot another i, Bob Cratchit!'

Scrooge was better than his word. He did it all, and infinitely more; and to Tiny Tim, who did NOT die,[2] he was a second father. He became as good a friend, as good a master, and as good a man, as the good old city knew, or any other good old city, town, or borough, in the good old world. Some people laughed to see the alteration in him, but his own heart laughed: and that was quite enough for him.

He had no further intercourse with Spirits, but lived in that respect upon the Total Abstinence Principle, ever afterwards; *and it was always said of him, that he knew how to keep Christmas well, if any man alive possessed the knowledge. May that be truly said of us, and all of us! And so, as Tiny Tim observed, God Bless Us, Every One!*[3]

[1] Stage-direction *Quick on*—needed because the *1843* text is here interrupted by an engraving.

[2] NOT in small capitals in *1843*. At this point, in one of the Charity Readings, 'a universal feeling of joy seemed to pervade the whole assembly, who rising spontaneously, greeted the renowned and popular author with a tremendous burst of cheering' (*Sheffield Daily Telegraph*, 11 December 1855). Dickens reported this 'most prodigious shout and roll of thunder' exultantly, to a friend (*N*, ii, 715).

[3] At the end of his American début (Boston, 2 December 1867), with this item, 'a dead silence seemed to prevail—a sort of public sigh as it were—only to be broken by cheers and calls, the most enthusiastic and uproarious, causing Mr. Dickens to break through his rule, and again presenting himself before his audience, to bow his acknowledgements' (Dolby, p. 174). Three weeks later, on Christmas Eve, he again read the *Carol* in Boston, with another kind of gratifying result: a local industrialist was so moved by hearing this reading that, next day, he changed his firm's practice of working on Christmas Day, and next year began the custom of giving a turkey to every employee (Gladys Storey, *Dickens and Daughter* (1939), pp. 120–1, quoted in the 1971 facsimile of the *Carol* Reading, pp. 205–6).

THE CHIMES

DICKENS's two pre-publication readings of *The Chimes* in December 1844, to groups of distinguished friends, were, wrote John Forster, 'the germ of those readings to larger audiences by which, as much as by his books, the world knew him in his later life' (*Life*, pp. 355–6, 363–4), so it was not surprising that this item was the second to appear in his professional career, on 6 May 1858. He was, however, 'a little afraid of it', finding difficulty in commanding 'sufficient composure at some of the more affecting parts, to project them with the necessary force, the requisite distance', and despite his reporting that it was a 'tremendous success' with audiences ('the effect was amazingly strong') it was soon dropped from his repertoire. '*Very* dramatic, but very melancholy on the whole', he came to think it. A decade later, regarding it as 'a little dismal', but still thinking he could 'make a good thing' of it, he 'shortened and brightened it as much as possible', but then gave only one performance of it in his Farewells.

In 1858 he sometimes prefaced this Reading by acknowledging that *The Chimes* very much reflected those social ills of 1844 which had provoked him to 'the utterance of a few earnest words', for which, happily, there was now 'less direct need'. In his 1868 revision, he deleted Will Fern's big speeches of social protest. Also he drastically reduced the supernatural machinery of the Bells and their Goblins, and the high-aspiring rhetorical passages about 'the Sea of Thought', 'the Sea of Time', and suchlike. The 1868 version (*Berg*), here reprinted, was set up from a (lost) earlier copy, Dickens's double underlinings in which were set by the 1868 printer in small capitals (here printed in italics).

Dickens's evocation of the domestic sentiments in this item was admired, as was his rendering of the pathos of Lilian's situation in the third Quarter, and Trotty Veck's appearances produced 'more tears and laughter combined than anything within the whole range of the acted drama' (*Critic*, 4 September 1858). 'To me it was a revelation', one spectator recalled nearly fifty years later:

> The man's beautiful, sympathetic voice, the wonderfully expressive eyes, his marvellous eloquence, his magnetic presence seemed to throw me under a spell . . . The pathos moved [the huge audience in the Manchester Free Trade Hall] to tears, the humour stirred them to roars of laughter. There were no accessories of music or scenery, simply one man at a reading-desk; but what a man! What a gift to be able to charm and sway a multitude! Sometimes you could have heard a pin drop, at others the roof seemed rent with the roars of the people as they gave vent to their strained feelings. And when it came to the peroration there was a silence which was almost painful, even a woman's sob here and there only served to intensify it . . . Gently, slowly the book was closed, and the solitary figure

seemed to glide from the stage, yet the vast audience remained silent—for hours; it was only seconds, but the seconds seemed hours. Then the people let themselves go; they had the weary man back, and they thundered their approval. He stood there slowly bowing, the tears of heartfelt emotion running down his pale cheeks. I passed out into the frosty night. I was a dreamer; I was dreaming dreams. Charles Dickens had carved his name on my heart. ('Dick Donovan' [J. E. Preston Murdock], *Pages from an Adventurous Life* (1907), pp. 53–5)

The Chimes

FIRST PART

* HIGH UP in the steeple of an old church, far above the town and far below the clouds, dwelt the Chimes I tell of.

Old Chimes,* but not speechless. They had clear, loud, lusty, sounding voices; and they rang out far and wide—'beating all other bells to fits,' as *Toby Veck*[1] said; for though they chose to call him Trotty Veck, his name was Toby. And I take my stand by Toby Veck, though he *did* stand all day long in all weathers, outside the church-door. In fact he was a ticket-porter, Toby Veck, and waited there for jobs. *

They called him Trotty from his pace, which meant speed if it didn't make it. He could have walked faster than he trotted; but rob him of his trot, and Toby would have taken to his bed and died. A weak, small, spare old man, he was a very Hercules, this Toby, in his strong intentions.

One day—*one New Year's Eve*[2]—Toby was trotting up and down before the Church, when *The Chimes*, his old daily companions, struck Twelve at Noon.

'Dinner-time, eh! Ah! * There's nothing more regular in its coming round than dinner-time, and there's nothing less regular in its coming round than dinner. I wonder whether it would be worth any gentleman's while, now, to buy that observation for the Papers; or the Parliament!'[3]

Toby was only joking with himself.

'Why! Lord! The Papers is full of observations as it is; and so's the Parliament. Here's last week's paper, now; full of observations! Full of observations! I like to know the news as well as any man; but it almost goes against the grain with me to read a paper now. It frightens me, almost. *I don't know what we poor people are coming to, our characters is so bad.*[4] Lord send we may be coming to something better in the New Year nigh upon us!'

'Father, father!'

Toby didn't hear.

'It seems as if we can't go right, or do right, or be righted. I hadn't

[1] *Toby Veck* in small capitals in *Berg*; *did* (in the next sentence) italic in *Berg* and in *1844*.

[2] Like *The Chimes* later in this sentence, this is doubly underlined in the manuscript insertion in *Berg*.

[3] Trotty spoke in 'a voice modified from Bob Cratchit's' (*Critic*, 4 September 1858, p. 538).

[4] Small capitals in *Berg*.

much schooling, myself, when I was young; and I can't make out whether we have any business on the face of the earth, or not. Sometimes I think we must have a little; and sometimes I think we must be intruding. I get so puzzled sometimes that I am not even able to make up my mind *whether there is any good at all in us, or whether we are born bad.*[1] We seem to do dreadful things; we seem to give a deal of trouble; we are always being complained of and guarded against. One way or another, we fill the papers. Talk of a New Year! Supposing it should really be that we have no right to a New Year—supposing we really *are*[2] intruding——'

'Why, father, father!'

Toby heard the pleasant voice, this time; started; stopped; and found himself face to face with his own child, and looking into her eyes.

Bright eyes they were. Eyes that would bear a world of looking in. Eyes that were beautiful and true, and beaming with Hope. With Hope so young and fresh; with Hope so buoyant, vigorous, and bright, despite the twenty years of work and poverty on which they had looked; that they became a Voice to Trotty Veck, and said: 'I think we have some business here—a little!'

Trotty kissed the lips belonging to the eyes, and squeezed the blooming face between his hands.

'Why Pet. What's to-do? I didn't expect you to-day, Meg.'

'Neither did I expect to come, father. But here I am! And not alone!'

'Why you don't mean to say,' *looking curiously at a covered basket which she carried in her hand,* 'that you have brought——'

'Smell it, father dear. Only smell it! Now. What's that?'

'Why, it's hot!'

'It's burning hot! It's scalding hot! But what is it, father? Come! You must guess what it is. I can't think of taking it out, till you guess what it is.' *

'Ah! It's very nice. It an't—I suppose it an't Polonies?'[3]

'No, no, no! Nothing like Polonies!'

'No. It's—it's mellower than Polonies. It's too decided for Trotters. Liver? No. There's a mildness about it that don't answer to liver. Pettitoes? No. It an't faint enough for pettitoes. It wants the stringiness of Cocks' heads. And I know it an't sausages. I'll tell you what it is. No, it is n't, neither. Why, what am I thinking of! I shall forget my own name next. It's tripe!'

Tripe it was; and Meg protested he should say, in half a minute more, it was the best tripe ever stewed.

[1] Small capitals in *Berg*.

[2] Italic in *Berg* and in *1844*.

[3] Much 'by-play' here by 'the Humorist . . . when he syllabled, with watering lips, guess after guess at the half-opened basket. "It ain't—I suppose it ain't polonies? [sniffing] . . .' (Kent, p. 165).

'And so I'll lay the cloth at once, father; for I have brought the tripe in a basin, and tied the basin up in a pocket handkerchief; and if I like to spread that for a cloth, and call it a cloth, there's no law to prevent me; is there, father?'

'Not as I knows of, my dear. But they're always a bringing up some new law or other.'*

'Make haste, father, for there's a hot potato besides, and half a pint of fresh-drawn beer in a bottle. Where will you dine, father? On the Post, or on the Steps? How grand we are. Two places to choose from!'

'The Steps to-day, my Pet. Steps in dry weather. Post in wet. There 's a greater conveniency in the Steps at all times, because of the sitting down; but they 're rheumatic in the damp.' *

As he was stooping to sit down, the Chimes rang.

'Amen!'

'Amen to the Bells, father?'

'They broke in like a grace, my dear. They'd say a good one, I am sure, if they could. For many's the kind thing they say to me. Why bless you, my dear,' *pointing at the tower with his fork*, 'how often have I heard them bells say "*Toby Veck*, *Toby Veck*, *keep a good heart*, *Toby*! *Toby Veck*, *Toby Veck*, *keep a good heart*, *Toby*!"[1] Have I heard 'em say it, a million times? More! When things is very bad indeed, then it's "*Toby Veck*, *Toby Veck*, *job coming soon*, *Toby*! *Toby Veck*, *Toby Veck*, *job coming soon*, *Toby*!" that way. * But Lord forgive me! My love! Meg! why did n't you tell me what a beast I was? Sitting here, cramming, and stuffing, and gorging myself; and you before me there, never so much as breaking your precious fast——'

'But I have broken it, father, all to bits. I have had my dinner, father. And if you 'll go on with yours I'll tell you how and where; and how your dinner came to be brought; and—and something else besides.'

So Trotty took up his knife and fork again.

'I had my dinner, father, with—with Richard. His dinner-time was early; and as he brought his dinner with him when he came to see me, we —we had it together, father.'

'Oh!'

'And Richard says, father——'

'What does Richard say, Meg?'

'Richard says, father——'

'Richard's a long time saying it.'

'He says then, father, another year is nearly gone, and where is the use of our waiting on from year to year, when it is so unlikely we shall ever be better off than we are now? He says we are poor now, father, and we shall be poor then; but we are young now, and years will make us old

[1] This and the following 'speech' of the Bells printed in small capitals in *Berg*.

before we know it. And how hard, father, to grow old, and die, and think we might have cheered and helped each other! How hard in all our lives to love each other; and to grieve, apart, to see each other working, changing, growing old and grey. So Richard says, father; as his work was yesterday made certain for some time to come, and as I love him and have loved him full three years—ah! longer than that, if he knew it!—will I marry him tomorrow—New Year's Day; the best and happiest day, he says, in the whole year. And he said so much, that I said I 'd come and talk to you, father. And as they paid the money for that work of mine this morning, and as you have fared very poorly for a whole week, and as I could n't help wishing there should be something to make this day a sort of holiday to you as well as a dear and happy day to me, father, I made a little treat and brought it to surprise you.'

'And see how he leaves it cooling on the step!'

The voice of this same Richard, looking down upon them with a face as glowing as the iron on which his stout sledge-hammer daily rang. A handsome, well-made, powerful youngster; with eyes that sparkled like the red-hot droppings from his furnace fire. *

Trotty reached up his hand to Richard, when the house door opened without any warning, and a footman very nearly put his foot in the tripe.

'Out of the vays here, will you! You must always go and be a settin on our steps, must you! You can't go and give a turn to none of the neighbours never, can't you! *Will*[1] you clear the road or won't you?'

Strictly speaking, the last question was irrelevant, because they had already done it.

'What's the matter, what's the matter!' *said the gentleman for whom the door was opened: coming out of the house.* 'What's the matter. What's the matter?'

'You 're always a being begged, and prayed, upon your bended knees you are, to let our doorsteps be. Why don't you let 'em be? CAN'T you let 'em be?'[2]

'There! That'll do, that'll do! Halloa there! Porter! Come here. What's that? Your dinner?'

'Yes, sir,' *leaving it behind him in a corner.*

'Don't leave it there. Bring it here, bring it here. So! This is your dinner, is it?'

'Yes, sir,' *looking, with a fixed eye and watery mouth, at the piece of tripe he had reserved for a last delicious tit-bit; which the gentleman was now turning over and over on the end of his fork.*

[1] Italic in *Berg* and in *1844*.

[2] CAN'T printed in capitals in *1844* as well as in *Berg*. Kent (p. 166), quoting this speech, notes that 'Nothing more was seen or heard of that footman, and yet in the utterance of those few words of his the individuality of the man somehow was thoroughly realized . . . he stood palpable there before us.'

Two other gentlemen had come out with him.†

He called to the first one by the name of Filer; and they both drew near together. Mr. Filer being exceedingly short-sighted, was obliged to go so close to the remnant of Toby's dinner before he could make out what it was, that Toby's heart leaped up into his mouth. But Mr. Filer did n't eat it.

'This is a description of animal food, *Alderman Cute*,'[1] *making little punches in it, with a pencil-case*, 'commonly known to the labouring population of this country, by the name of tripe. But who eats tripe? Tripe is without an exception the most wasteful article of consumption that the markets of this country can by possibility produce. The loss upon a pound of tripe has been found to be, in the boiling, seven eighths of a fifth more than the loss upon a pound of any other animal substance whatever. Tripe is more expensive, properly understood, than the hothouse pineapple. Taking into account the number of animals slaughtered yearly within the bills of mortality alone; and forming a low estimate of the quantity of tripe which the carcases of those animals, reasonably well butchered, would yield; I find that the waste on that amount of tripe, if boiled, would victual a garrison of five hundred men for five months of thirty-one days each, and a February over. The Waste, the Waste!'

Trotty stood aghast. He seemed to have starved a garrison of five hundred men with his own hand.

'Who eats tripe? Who eats tripe?'

Trotty made a miserable bow.

'You do, do you? Then I'll tell you something. You snatch your tripe, my friend, out of the mouths of widows and orphans.'

'I hope not, Sir. I'd sooner die of want!'

'Divide the amount of tripe before-mentioned, Alderman, by the estimated number of existing widows and orphans, and the result will be one pennyweight of tripe to each. Not a grain is left for that man. Consequently, he 's a robber.'

Trotty was so shocked, that it gave him no concern to see the Alderman finish the tripe himself. It was a relief to get rid of it, anyhow.

'And what do you say?' asked the Alderman, jocosely, of his other friend. 'You have heard friend Filer. What do *you*[2] say?'

'What 's it possible to say? What *is* to be said? Who can take any interest in a fellow like this? Look at him! What an object! Look into Strutt's Costumes, and see what a Porter used to be in any of the good old English reigns. Ah! the good old times, the grand old times, the great old times!'

The gentleman did n't specify what particular times he alluded to. †

[1] *Alderman Cute* in small capitals in *Berg*; the phrase that follows is simply underlined.

[2] *you* (and *is* in the next sentence) italic in *Berg* and in *1844*.

'Now, you know,' *said the Alderman, addressing his two friends*, 'I am a plain man, and a practical man; and I go to work in a plain practical way. That's my way. There is not the least mystery or difficulty in dealing with this sort of people if you only understand 'em, and can talk to 'em in their own manner. Now, you Porter! Don't you ever tell me, or anybody else my friend, that you have n't always enough to eat, and of the best; because I know better. I have tasted your tripe, you know, and you can't "chaff" me. You understand what "chaff" means, eh? That's the right word is n't it? Ha, ha, ha! Lord bless you, it's the easiest thing on earth to deal with this sort of people, if you only understand 'em. You see my friend, there's a great deal of nonsense talked about Want—"hard up," you know: that's the phrase is n't it?—and I intend to Put it Down. That's all! Lord bless you, you may Put Down anything among this sort of people, if you only know the way to set about it!'

Trotty took Meg's hand and drew it through his own.

'Your daughter, eh? Where's her mother?'

'Dead!'

'Oh! and you're making love to her, are you? you young smith?'

'Yes. And we are going to be married on New Year's Day.'

'What do you mean! Married?' said Mr. Filer.

'Why, yes, we're thinking of it, Master. We're rather in a hurry you see, in case it should be Put Down first.'

'Ah! Put *that*[1] down indeed, Alderman, and you'll do something. Married! Married!! A man may live to be as old as Methusaleh, and may labour all his life for the benefit of such people as those; and may heap up facts on figures, facts on figures, facts on figures, mountains high and dry; and he can no more hope to persuade 'em that they have no right or business to be married, than he can hope to persuade 'em that they have no earthly right or business to be born. And *that* we know they have n't. We reduced it to a mathematical certainty long ago.'

Alderman Cute laid his right forefinger on the side of his nose, as much as to say to both his friends, 'Observe me, will you? Keep your eye on the practical man!'—and called Meg to him.

'Come here, my girl! * You are going to be married, you say. Rather unbecoming and indelicate in one of your sex! But never mind that. After you are married, you'll quarrel with your husband, and come to be a distressed wife. You may think not: but you will, because I tell you so. Now I give you fair warning, that I have made up my mind to Put distressed wives Down. So don't be brought before me. You'll have children—boys. Those boys will grow up bad of course, and run wild in the streets, without shoes and stockings. Mind, my young friend! I am determined to Put boys without shoes and stockings, Down. Perhaps

[1] *that* here, and *that* again at the end of the paragraph, italic in *Berg* and in *1844*.

your husband will die young (most likely) and leave you with a baby. Then you'll be turned out of doors, and wander up and down the streets. Now don't wander near me, my dear, for I am resolved to Put all wandering mothers Down. And, above all, don't you attempt to drown yourself, or hang yourself, for I have made up my mind to Put all suicide Down. As for you, you dull dog, what are you thinking of being married for? What do you want to be married for, you silly fellow? If I was a fine young, strapping chap like you, I should be ashamed of being milksop enough to pin myself to a woman's apron-strings! Why, she'll be an old woman before you're a middle-aged man! And a pretty figure you'll cut then, with a draggle-tailed wife and a crowd of squalling children crying after you wherever you go! There! Go along with you and repent. †—Porter, don't you go. As you happen to be here, you shall carry a letter for me. Can you be quick? You're an old man.—How old are you?'

'I'm over sixty-eight, Sir,' said Toby.[1]

'Oh! This man's a great deal past the average age, you know,' cried Mr. Filer, *breaking in as if his patience would bear some trying, but this really was carrying matters a little too far*.

'Yes: I feel I'm intruding, Sir,' said Toby. 'I—I misdoubted it this morning!'

The Alderman cut him short by giving him the letter from his pocket. Toby would have got a shilling too; but Mr. Filer clearly showing that in that case he would rob a certain given number of persons of ninepence-halfpenny a-piece, he only got sixpence; and thought himself very well off to get that.[2] *

The letter was addressed to a great man in the great district of town. The greatest district of the town. It must have been the greatest district of the town, because it was commonly called The World by its inhabitants. *

'*Put 'em down, put 'em down, Facts and Figures, facts and figures, Good old Times, good old times. Put 'em down, put 'em down.*'[3] The unfaithful and unfeeling Chimes went to that measure, and Toby's trot went to that measure, and neither Chimes nor Trot would go to any other burden.

But even that burden brought him, in due time, to the end of his journey. To the mansion of Sir Joseph Bowley, Member of Parliament.

The door was opened by a Porter. Such a Porter![4] When he had found

[1] *1844* has 'over sixty'—a characteristic example of the raising of numerals in the Readings.

[2] The rest of the 'First Quarter' is here omitted, and a much abbreviated version of the opening of the 'Second Quarter' follows.

[3] These sentences printed in small capitals in *Berg*.

[4] This porter (Tugby) was 'a platform creation of the highest dramatic order, built up out of a few lines in the book, which an ordinary reader would pass by' (*Critic*, 4 September 1858, p. 537). His voice was 'a fat whisper' (Kent, p. 167, quoting a phrase from the novel, here omitted).

his voice—which it took him some time to do, for it was a long way off, and hidden under a load of meat—he said:

'Who's it from?'

Toby told him.

'You're to take it in, yourself. Everything goes straight in, on the last day of the old year.'

Toby wiped his feet (which were quite dry already) and took the way pointed out to him; observing as he went that it was an awfully grand house. Knocking at the room door, he was told to enter from within; and, doing so, found himself in a library, where, at a table strewn with files and papers, were a stately lady in a bonnet; and a not very stately gentleman in black who wrote from dictation; while another, and an older, and a much statelier gentleman walked up and down, with one hand in his breast.

'What is this?' said the last-named gentleman. 'Mr. Fish, will you have the goodness to attend?'

Mr. Fish begged pardon, and taking the letter from Toby, handed it, with great respect.

'From Alderman Cute, Sir Joseph.'

'Is this all? Have you nothing else, Porter? You have no bill or demand upon me (my name is Bowley, Sir Joseph Bowley) of any kind from anybody, have you? I allow nothing to be carried into the New Year. Every description of account is settled in this house at the close of the old one. So that if death was to—to—sever the cord of existence—my affairs would be found, I hope, in a state of preparation.'

'My dear Sir Joseph! How shocking!'

'My lady Bowley, at this season of the year we should think of—of—ourselves. We should look into our—our —accounts. We should feel that every return of so eventful a period in human transactions, involves matters of deep moment between a man and his—and his banker.' *

'Ah! you are the Poor Man's Friend, you know, Sir Joseph.'

'I *am*[1] the Poor Man's Friend.'

'Bless him for a noble gentleman!'

'I don't agree with Cute here, for instance. I don't agree with the Filer party. I don't agree with any party. My friend the Poor Man, has no business with anything of that sort, and nothing of that sort has any business with him. My friend the Poor Man, in my district, is my business. Your only business, my good fellow,' looking abstractedly at Toby; 'your only business in life is with me. You need n't trouble yourself to think about anything. I will think for you; I know what is good for you; I am your perpetual parent. Such is the dispensation of an all-wise Providence! Now, the design of your creation is: not that you should swill, and guzzle, and associate your enjoyments, brutally, with food'—

[1] Italic in *Berg* and in *1844*.

Toby thought remorsefully of the tripe[1]—'but that you should feel the Dignity of Labor; go forth erect into the cheerful morning air, and—and stop there.' *

'Ah! you have a thankful family, Sir Joseph!'

'My lady, ingratitude is known to be the sin of that class. I expect no other return.'

'*Ah*! *Born bad*! Nothing melts us!' *

Sir Joseph opened the Alderman's letter.

'Very polite and attentive, I am sure! My lady, the Alderman is so obliging as to inquire whether it will be agreeable to me to have *Will Fern* put down.'

'*Most*[2] agreeable! The worst man among them! He has been committing a robbery, I hope?'

'Why, no; not quite. Very near. Not quite. He came up to London, it seems, to look for employment (trying to better himself—that's his story), and being found at night asleep in a shed, was taken into custody and carried next morning before the Alderman. The Alderman observes (very properly) that he is determined to put this sort of thing down; and that if it will be agreeable to me to have Will Fern put down, he will be happy to begin with him.'

'Let him be made an example of, by all means,' returned the lady. 'Last winter, when I introduced pinking and eyelet-holeing among the men and boys in the village, as a nice evening employment, and had the lines—

Oh let us love our occupations,
Bless the squire and his relations,
Live upon our daily rations,
And always know our proper stations—

set to music on the new system, for them to sing the while; this very Fern —I see him now—touched that hat of his, and said, "I humbly ask your pardon my lady, but *an't* I something different from a great girl?" Make an example of him.'

'Hem! Mr. Fish, if you'll have the goodness to attend'——

Mr. Fish seized his pen, and wrote from Sir Joseph's dictation.

'Private. My dear Sir. I am very much indebted to you for your courtesy in the matter of the man William Fern, of whom, I regret to add, I can say nothing favourable. I have uniformly considered myself in the light of his Friend and Father, but have been repaid (a common case, I grieve to say) with ingratitude, and constant opposition to my plans. He is a turbulent and rebellious spirit. His character will not bear investigation. Nothing will persuade him to be happy when he might. Under these circumstances, it appears to me, I own, that when he comes before you

[1] This and the next two emphases (*Ah*! *Born bad*! and *Will Fern*) in small capitals in *Berg*.

[2] *Most*, and the next emphasis (*an't* at the end of the next paragraph but one) italic in *Berg* and in *1844*.

again his committal for some short term as a Vagabond, would be a service to society. And I am,' and so forth.

'With my compliments and thanks, Porter. Stop! You have heard, perhaps, certain remarks into which I have been led respecting the solemn period of time at which we have arrived, and the duty imposed upon us of settling our affairs, and being prepared. Now, my friend, can you lay your hand upon your heart, and say, that you also have made preparation for a New Year?'

'I am afraid, Sir, that I am a—a—little behind-hand with the world.'

'Behind-hand with the world!'

'I am afraid, Sir, that there's a matter of ten or twelve shillings owing to Mrs. Chickenstalker.'

'To Mrs. Chickenstalker!'

'A Shop, Sir, in the general line. Also I'm fearful there 's a—a little money owing on account of rent. A very little, Sir. It ought n't to be owing, I know, but we have been hard put to it, indeed!'

Sir Joseph looked at his lady, and at Mr. Fish, and at Trotty, one after another, twice all round. He then made a despondent gesture with both hands at once, as if he gave the thing up altogether.

'There! Take the letter. Take the letter! Take the letter, take the letter!' He had nothing for it but to make his bow and leave the house. In the street, he pulled his old hat down upon his head, to hide the grief he felt at getting no hold on the New Year, anywhere. *

Unfaithful and unfeeling Chimes! They would n't cheer up Toby now. '*Facts and figures, facts and figures. Good old times, good old times. Born bad, born bad! Put 'em down, put 'em down*!'[1] He could hear the Chimes ring nothing better.

He discharged himself of his commission, and set off trotting homeward. But what with his pace, and what with his hat over his eyes, he soon trotted against somebody.

'I beg your pardon, I'm sure! I hope I have n't hurt you!'

The man against whom he had run; a sun-browned country-looking man, replied:

'No, friend. You have not hurt me.'

'Nor the child, I hope?'

'Nor yet the child. I thank you kindly.'

He glanced at a little girl he carried in his arms, asleep. *

'You can tell me, perhaps—and if you can I am sure you will, and I'd rather ask you than another—where Alderman Cute lives.'

'It's impossible your name's Fern! Will Fern!'

'That's my name.'

'Why then, for Heaven's sake don't go to him! Don't go to him! He'll

[1] Small capitals in *Berg*.

put you down as sure as ever you were born. Here! come up this alley, and I'll tell you what I mean. Don't go to *him*.'[1]

His new acquaintance looked as if he thought him mad; but he bore him company. When they were shrouded from observation, Trotty told him what he knew, and what character he had received, and all about it.

The countryman did not contradict. He nodded his head now and then; and threw back his hat, and passed his hand over a brow, *where every furrow he had ploughed seemed to have set its image in little*.[2]

'It's true enough in the main, master. I could sift grain from husk here and there, but let it be as 't is. What odds? I have gone against his plans; to my misfortun'. I can't help it; I should do the like to-morrow. As to character, them gentlefolks will search and search, and pry and pry, and have it as free from spot or speck in us, afore they'll help us to a dry good word!—Well! I hope they don't lose good opinion as easy as we do, or their lives is strict indeed. For myself, master, I never took with that hand what was n't my own; and never held it back from work, however hard, or poorly paid. Whoever can deny it, let him chop it off! But when work won't maintain me like a human creetur; when my living is so bad, that I am Hungry, out of doors and in; then I say to the gentlefolks "Keep away from me! Let my cottage be. My doors is dark enough without your darkening of 'em more. We've nought to do with one another. I'm best let alone!" '

Seeing that the child was awake, he checked himself to say a word or two of prattle in her ear, and stand her on the ground beside him. Then, while she hung about his dusty leg, he said to Trotty:

'I'm not a cross-grained man by natur', I believe; and easy satisfied, I'm sure. I bear no ill will against none of 'em; but I've got a bad name this way, and I'm not likely, I'm afeard, to get a better. 'T an't lawful to be out of sorts, and I AM[3] out of sorts, though God knows I'd sooner bear a cheerful spirit if I could. Well! I don't know as this Alderman could hurt *me* much by sending of me to gaol; but without a friend to speak a word for me, he might do it; and you see—!' *pointing downward with his finger, at the child*.

'She has a beautiful face.'

'Why yes! I've thought so, many times. I've thought so, when my hearth was very cold, and cupboard very bare. I thought so t'other night, when we were taken like two thieves. But they—they shouldn't try the little face too often, should they, Lilian? That's hardly fair upon a man!'

'Is your wife alive?'

'I never had one. She's little Lilian—my brother's child: a orphan. Nine year old. They'd have took care on her, the Union; *eight and twenty*

[1] Italic in *Berg* and in *1844*. [2] Small capitals in *Berg*.
[3] AM in small capitals, and *me* (just below) in italics, in *1844* as well as in *Berg*. The final words of the paragraph are in small capitals in *Berg* only.

mile away from where we live[1] (as they took care of my old father when he could n't work no more, though he did n't trouble 'em long); but I took her instead, and she's lived with me ever since. Her mother had a friend once, in London here. We are trying to find her, and to find work too; but it's a large place. Never mind. More room for us to walk about in, Lilly!'

Meeting the child's eyes with a smile which melted Toby more than tears, he shook him by the hand.

'I don't so much as know your name; but I've opened my heart free to you, for I'm thankful to you. I'll take your advice. And to-morrow Lilly and me will try whether there's better fortun' to be met with, somewheres near London. Good night. A Happy New Year!'

'Stay! Stay! The New Year never can be happy to me, if we part like this. The New Year never can be happy to me, if I see the child and you go wandering away, you don't know where, without a shelter for your heads. Come home with me! I'm a poor man, living in a poor place; but I can give you lodging for one night and never miss it. Come home with me! Here! I'll take her! A pretty one! I'd carry twenty times her weight, and never know I'd got it. Why, she's as light as a feather.—*Here*[2] we are, and here we go!—Round this first turning to the right, Uncle Will, and past the pump, and sharp off up the passage to the left, right opposite the public-house.—*Here* we are, and here we go.—Cross over, Uncle Will, and mind the kidney pieman at the corner!—*Here* we are, and here we go! —Down the Mews here, Uncle Will, and stop at the black door, with "T. Veck, Ticket-Porter," wrote upon a board; and—*here* we are, and here we go—and here we are indeed, my precious Meg, surprising of you!'

He set the child down before his daughter in the middle of the floor. The little visitor looked once at Meg; and doubting nothing in that face, but trusting everything she saw there, ran into her arms.

'Here we are and here we go!' *running round the room, and choking audibly.* '*Here*! *Uncle Will*! Why don't you come to the fire?—Oh here we are and here we go!—Meg, my precious darling, where's the kettle?—Here it is and here it goes, and it 'll bile in no time!'[3]

Trotty really had picked up the kettle somewhere or other, and now put it on the fire: while Meg, seating the child in a warm corner, kneeled down on the ground before her, and pulled off her shoes, and dried her wet feet.

'Why father! You're crazy to-night, I think. I don't know what the Bells would say to that.—Poor little feet. How cold they are! Why father! Good gracious me! He's crazy! He's put the dear child's bonnet on the kettle, and hung the lid behind the door!'

[1] Small capitals in *Berg*.

[2] *Here* throughout this paragraph printed in small capitals in *Berg*.

[3] 'It was in the touching scenes of home affection [such as this], . . . that both as an author and a highly cultivated elocutionist, he was most unquestionably at home' (*Saunders' News-letter*, 26 August 1858).

'I didn't go to do it, my love. Meg, my dear!'

Behind the chair of their male visitor, he was holding up the sixpence he had earned.

'I see, my dear, as I was coming in, half an ounce of tea lying somewhere on the stairs; and I'm pretty sure there was a bit of bacon too. As I don't remember where it was, exactly, I'll go myself and try to find 'em.'

With this inscrutable artifice, Toby withdrew to purchase the viands at Mrs. Chickenstalker's; and presently came back, pretending that he had not been able to find them, at first, in the dark.

'But here they are at last,' *said Trotty, setting out the tea-things*, 'all correct! I was pretty sure it was tea, and a rasher. So it is. Meg, my Pet, if you'll just make the tea, while your unworthy father toasts the bacon, we shall be ready, immediate. It's a curious circumstance,' said Trotty, proceeding in his cookery, 'curious, but well known to my friends, that I never care, myself, for rashers, nor for tea. I like to see other people enjoy 'em,' speaking very loud, to impress the fact upon his guest, 'but to me, as food, they 're disagreeable.' *

No. Trotty's occupation was to see Will Fern and Lilian eat and drink. And so was Meg's. *

'Now, I'll tell you what,' *said Trotty after tea*. 'The little one, she sleeps with Meg, I know. * Will Fern, you come along with me. You're tired to death, and broken down for want of rest. You come along with me. I'll show you where you lie. It 's not much of a place: only a loft: but there's plenty of sweet hay up there belonging to a neighbour; and it's as clean, as hands and Meg can make it. Cheer up! Don't give way. A new heart for a New Year, always!'

It was some short time before the foolish little old fellow, left to himself, could compose himself to mend the fire, and draw his chair to the warm hearth. But when he had done so, and had trimmed the light, he took his newspaper from his pocket, and began to read. Carelessly at first, and skimming up and down the columns; but with an earnest and a sad attention, very soon.

For this same dreaded paper re-directed Trotty's thoughts into the channel they had taken all that day. His interest in the two wanderers had set him on another course of thinking, and a happier one; but being alone again, and reading of the crimes and violences of the people, he relapsed into his former train.

In this mood, he came to an account (and it was not the first he had ever read) of a woman who had laid her desperate hands not only on her own life; but on that of her young child.[1] A crime so terrible, and so revolting to his soul, dilated with the love of Meg, that he let the journal drop.

'Unnatural and cruel! Unnatural and cruel! None but people who were bad at heart: born bad: who had no business on the earth: could do such

[1] Small capitals in *Berg*.

deeds. It's too true, all I've heard to-day in the Chimes, and out of them; it's too just, too full of proof. Put us Down. We're Bad!'

The Chimes took up the words so suddenly that the Bells seemed to strike him out of his chair.

And what was that, they said?

'*Toby Veck*, *Toby Veck*, *waiting for you*, *Toby*! *Toby Veck*, *Toby Veck*, *waiting for you*, *Toby*! *Toby Veck*, *Toby Veck*, *door open wide*, *Toby*; *Toby Veck*, *Toby Veck*, *door open wide*, *Toby*![1] Haunt his slumbers, Haunt his slumbers! Come and see us, Come and see us!——'

'Meg,' *tapping at her door*. 'Do you hear anything?'

'I hear the Bells, father. Surely they're very loud to-night.'

He resumed his seat by the fire, and once more listened. *He fell off into a doze*; *then roused himself*. It was impossible to bear the Bells; their energy was dreadful. †

So he slipped out into the street * and went in to the Bell-Tower, feeling his way.[2] It was very dark. And very quiet, for the Chimes were silent now. He groped his way, and went up, until, ascending through the floor, and pausing with his head just raised above its beams, he came among the Bells. * Then did he see in every Bell a bearded figure, of the bulk and stature of the Bell—incomprehensibly, a figure and the Bell. Gigantic, grave, and darkly watchful of him, as he stood rooted to the ground. Mysterious and awful figures! Resting on nothing; poised in the night air of the tower, motionless and shadowy! *

The Great Bell, or the Goblin of the Great Bell, spoke. 'What visitor is this!'

'I thought my name was called by the Chimes. I hardly know why I am here, or how I came here. I have listened to the Chimes these many years. They have cheered me often.'

'And you have thanked them?'

'A thousand times!'

'How?'

'I am a poor man, and could only thank them in words.'

'Have you never done us wrong in words?' *

'I never did so, to my knowledge, Sir; it was quite by accident if I did. I would n't go to do it, I'm sure.' *

'Who hears in us, the Chimes, one note bespeaking disregard, or stern regard, of any hope, or joy, or pain, or sorrow, of the many-sorrowed throng; who hears us make response to any creed that gauges human passions and affections, as it gauges the amount of miserable food on

[1] Small capitals in *Berg*. The remainder of the Bells' 'speech' was inserted, in manuscript, in *Berg*. Also printed in small capitals in *Berg* is the sentence below, *He fell off into a doze*; *then roused himself*.

[2] This paragraph is a drastically condensed version of the closing pages of the 'Second Quarter' and the opening of the 'Third Quarter'.

which humanity may pine and wither; does us wrong. That wrong you have done us! * Lastly, and most of all, who turns his back upon the fallen and disfigured of his kind; abandons them as Vile from the beginning; and does not trace with pitying eyes the unfenced precipice by which they fell from Good; who does this wrong to Heaven and Man, to Time and to Eternity. And you have done that wrong.'

'I hope not, spirits of the Bells. I should be sorry to do such wrong at any time, but most of all upon a New Year's Eve.'

'A New Year's Eve! Listen![1] The New Year is past—nine years ago. You ceased from among the living, nine years ago. You missed your foothold on the outside of this tower in the dark, and fell into the deep street, nine years ago. Your child is living. *Learn from her life, a living truth.*[2] *Learn from the creature dearest to your heart, how bad the Bad are born.* See every bud and leaf plucked one by one from the fairest stem, and know how bare and wretched it may become! *Follow her! To Desperation!*'

Each of the figures stretched its right arm forth, and pointed downward. And then, where the Figures had been, the Bells were.

SECOND PART

WITH a confused and stunned sensation of not being able to make himself seen, or heard—of being a mere shade without substance—not alive and yet sentient—Trotty Veck looked into a strange unearthly atmosphere before him.

In a poor, mean room; working at the same kind of embroidery which he had often, often, seen before her; Meg, his own dear daughter, was presented to his view.

Changed. Changed. The light of the clear eye, how dimmed. The bloom, how faded from the cheek.

She looked up from her work, at a companion. Following her eyes, the old man started back. For, in the woman grown, he recognized Lilian Fern. *

Hark. They were speaking!

'Meg, how often you raise your head from your work to look at me!'

'Are my looks so altered, that they frighten you?'

'Nay dear! But when you think I'm busy, and don't see you, you look so anxious and so doubtful, that I hardly like to raise my eyes. There is little cause for smiling, in this hard and toilsome life of ours; but you were once so cheerful.'

[1] The next two paragraphs are a condensed and re-arranged version of the 1844 text (Oxford Illustrated edition, pp. 124–5).

[2] This sentence in small capitals in *Berg*; the next sentence is printed in italic, but with *how* in small capitals. The last four words of this speech are also printed in small capitals.

'Am I not now! Do *I*[1] make our weary life more weary to you, Lilian!'

'You have been the only thing that made it life, sometimes the only thing that made me care to live so, Meg. Such work, such work! So many hours, so many days, so many nights of hopeless, cheerless, never-ending work—not to heap up riches; but to earn bare bread! Oh Meg, Meg! How can the cruel world go round, and bear to look upon such lives!'

'Lilly! Why Lilly! You! So pretty and so young!'

'Oh Meg! *The worst of all, the worst of all! Strike me old, Meg! Wither me and shrivel me, and free me, from the dreadful thoughts that tempt me in my youth!*'[2] *

His former stunned sensation came on Trotty, and he saw nothing but mist. It cleared; he rubbed his eyes, and looked again.

His daughter was again seated at her work.[3] But in a poorer, meaner garret than before; *and with no Lilian by her side.*

The frame at which Lilian had worked, was put away upon a shelf and covered up. The chair in which Lilian had sat, was turned against the wall. *

A knock came at Margaret's door. She opened it. A man was on the threshold. A slouching, moody, drunken sloven: wasted by intemperance and vice: and with his matted hair and unshorn beard in wild disorder: *but with some traces on him, too, of having been a man of good proportion and good features in his youth. Trotty knew him. Richard.*[4]

'May I come in, Margaret?'

'Yes! Come in. Come in!' †

'Still at work, Margaret? You work late.'

'I generally do.'

'And early?'

'And early.'

'So she said. She said you never tired; or never owned that you tired. Not all the time you lived together. But I told you that, the last time I came. Margaret! What am I to do? She has been to me again!'[5]

'Again! Oh! does she think of me so often! Has she been again!'

'Twenty times again. Margaret, she haunts me. She comes behind me in the street, and thrusts it in my hand. I hear her foot upon the ashes when

[1] Italic in *Berg* and in *1844*.

[2] Small capitals in *Berg*. The whole of the Bowley Hall episode, which follows here in *1844*, is omitted in *Berg*. This episode included Will Fern's long protest speech.

[3] Small capitals in *Berg*; so is the phrase in the next sentence.

[4] All in small capitals in *Berg*.

[5] Richard's account of Lilian was (according to the *Clifton Chronicle*, 4 August 1858) 'the finest bit of the whole reading,' enacted with 'wonderful art.... You saw before you the wretched, starving, pitying, loving, upright Meg, and the haggard, wild and restless Richard....'

I'm at my work (ha, ha! *that*[1] an't often now), and before I can turn my head, her voice is in my ear, saying, "Richard, don't look round. For heaven's love, give her this!" She brings it where I live; she sends it in letters; she taps at the window and lays it on the sill. What *can*[2] I do? Look at it!'

'Hide it, hide it! When she comes again, tell her, Richard, that I love her in my soul. That I never lie down to sleep, but I bless her, and pray for her. That in my solitary work, I never cease to have her in my thoughts. That she is with me night and day. That if I died to-morrow, I would remember her with my last breath. But that I cannot look upon it!'

'I told her so. I told her so, as plain as words could speak. I've taken this gift back and left it at her door, a dozen times since then. But when she came at last, and stood before me, face to face, what could I do?'

'You saw her!'

'I saw her. There she stood: trembling! Says she, "How does she look, Richard? Does she ever speak of me? Richard, I have fallen very low; and you may guess how much I have suffered in having this sent back, when I can bear to bring it in my hand to you. But you loved her once, even in my memory, dearly."—(I suppose I did.—I did! That's neither here nor there now.)—"Oh Richard, if you have any memory for what is gone and lost, take it to her once more. Once more! Tell her how I laid my head upon your shoulder, *where her own head might have lain*,[3] and was so humble to you, Richard. Tell her that you looked into my face, and saw the beauty which she used to praise, all gone: all gone: and in its place, a poor, wan, hollow cheek, that she would weep to see."—You won't take it, Margaret?'

She shook her head, and motioned an entreaty to him to leave her.

'Good night, Margaret.'

'Good night!'

She sat down to her task, and plied it. Night, midnight. Still she worked. The Chimes rang half-past twelve; and there came a gentle knocking at the door. It opened.

She saw the entering figure; screamed its name; cried 'Lilian!'

It was swift, and fell upon its knees before her: clinging to her dress.

'Up, dear! Up! Lilian! My own dearest!'

'Never more, Meg; never more!'

'Sweet Lilian! Darling Lilian! Child of my heart—no mother's love can be more tender—lay your head upon my breast and let me raise you!'

'Never more, Meg. Never more! When I first looked into your face, *you*[4] kneeled before *me*. On my knees before *you*, let me die. Let it be here!'

[1] Small capitals in *Berg*.

[2] Italic in *Berg* and in *1844*. The 'it' referred to is a purse; Dickens deleted the text referring to it, as he could convey the sense of gesture.

[3] Small capitals in *Berg*. [4] *you . . . me . . . you* italic in *Berg*.

'You have come back. We will live together, work together, hope together, die together!'

'Ah! Kiss my lips, Meg; fold your arms about me; press me to your bosom; look kindly on me; but don't raise me. Let me see the last of your dear face upon my knees! His blessing on you, dearest love. He suffered her to sit beside His feet, and dry them with her hair. Oh Meg, His Mercy and Compassion!'

Her heart was broken, and she died in the encircling arms.

THIRD PART

SOME new remembrance of the figures in the Bells, some new remembrance of the ringing of the Chimes, some new knowledge—how conveyed to him he knew not—that more years had passed, and Trotty Veck, again without the power of being heard or seen, again looked on at mortal company.

Fat company, rosy-cheeked company, comfortable company. They were but two, but they were red enough for ten.[1] They sat before a bright fire, with a small low table between them. * The fire gleamed not only in the little room, and on the panes of window-glass in the door, and on the curtain half drawn across them, but in the little shop beyond. A little shop in the chandlery-way, or general line, quite crammed and choked with the abundance of its stock. * Trotty had small difficulty in recognising in the stout old lady, Mrs. Chickenstalker. * In Mrs. Chickenstalker's partner in the general line, and in the crooked and eccentric line of life, he recognized the former porter of Sir Joseph Bowley. *

'What sort of night is it, Anne?' inquired the former porter of Sir Joseph Bowley, stretching out his legs before the fire.

'Blowing and sleeting hard, and threatening snow. Dark. And very cold.'

'I'm glad to think we had muffins for tea, my dear. It's a sort of night that's meant for muffins. Likewise crumpets. Also Sally Lunns.'

'You 're in spirits, Tugby.'

(The firm was Tugby, late Chickenstalker.)

'No. Not particular. I'm a little elewated on accounts of being comfortable in-doors while it's such bad weather outside. * There's a customer, my love!'

Attentive to the rattling door, Mrs. Tugby had already risen.

'Now then. What's wanted? Oh! I beg your pardon, Sir, I'm sure. I did n't think it was you.'

She made this apology to a gentleman in black, who sat down astride on a table-beer barrel, and seemed to be some authorized medical attendant on the poor.

[1] 'A roar invariably greeted [this] remark' (Kent, p. 174).

'This is a bad business up-stairs, Mrs. Tugby. The man can't live.'

'Not our back-attic can't!'

'Your back-attic, Mr. Tugby, is coming down-stairs fast; and will be below the basement very soon.—The back-attic, Mr. Tugby, is Going.'

'Then,' *turning to his wife*, 'he must Go, you know, before he's Gone. It's the only subject that we've ever had a word upon, she and me, and look what it comes to! He's going to die here, after all. Going to die upon the premises. Going to die in our house!'

'And where should he have died, Tugby!'

'In the workhouse. What are workhouses made for?'

'Not for that. Not for that. Neither did I marry you for that. Don't you think it, Tugby. I won't have it. I won't allow it. I'd be separated first, and never see your face again. When my widow's name stood over that door, as it did for many, many years, I knew him, Richard, a handsome, steady, manly, independent youth; I knew her, Meg Veck, as the sweetest-looking girl eyes ever saw; and when I turn them out of house and home, may angels turn me out of Heaven. As they would! And serve me right!'

'Bless her! Bless her!'

'There's something interesting about the woman, even now. How did she come to marry him?'

'Why that is not the least cruel part of her story, Sir. You see they kept company, she and Richard, many year ago, and they were to have been married on a New Year's Day. But, somehow, Richard got it into his head, through what the gentlemen told him, that he might do better, and that he'd repent it, and the gentlemen frightened *her*,[1] and made her timid of his deserting her, and a good deal more. And in short, they lingered and lingered, and their trust in one another was broken, and so at last was the match. But never did woman grieve more truly for man, than she for Richard when he first went wrong.'

'Oh! he went wrong, did he?'

'Well, Sir, I don't know that he rightly understood himself, you see. I think his mind was troubled by their having broke with one another. He took to drinking, idling, bad companions. He lost his looks, his character, his health, his strength, his friends, his work, everything! This went on for years and years. At last, he was so cast down, and cast out, that no one would employ him. Applying for the hundredth time to one gentleman who had often and often tried him (he was a good workman to the very end); that gentleman, who knew his history, said, "I believe you are incorrigible; there is only one person in the world who has a chance of reclaiming you; ask me to trust you no more, until she tries to do it."

[1] Doubly underlined in *Berg*.

Something like that in his anger and vexation.—Well, Sir; Richard went to her, and made a prayer to her to save him.'

'And she—Don't distress yourself, Mrs Tugby.'

'She came to me that night to ask me about living here. "What he was once to me," she said, "is buried in a grave; side by side with what I was once to him. But I have thought of this; and I will make the trial." So they were married; and when they came home here, and I saw them, I hoped that such prophecies as parted them when they were young, may not often fulfil themselves as they did in this case, or I would n't be the makers of them for a Mine of Gold.'

'I suppose he used her ill?'

'I don't think he ever did that, Sir. He went on better for a short time; but his habits were too old and strong to be got rid of. There he has been lying now, these weeks and months. Between him *and her baby*,[1] she has not been able to do her old work; and by not being able to be regular, she has lost it, even if she could have done it. How they have lived, I hardly know!'

'*I*[2] know,' muttered Mr. Tugby; *looking at the till, and round the shop, and at his wife; and rolling his head with immense intelligence.* 'Like Fighting Cocks!'

He was interrupted by a cry from the upper story of the house. The gentleman ran up-stairs; Trotty floated up like mere air.

'*Follow her! Follow her! Follow her!*'[3] *He heard the ghostly voices in the Bells repeat their words.* 'Learn it from the creature dearest to your heart!'

It was over. It was over. The ruins of Richard cumbered this earth no more. And this was she, her father's pride and joy! This haggard, wretched woman, weeping by the bed, if it deserved that name, and pressing to her breast an infant. Who can tell how spare, how sickly, and how poor an infant? Who can tell how dear?

'Thank God!' cried Trotty, *holding up his folded hands.* '*Oh, God be thanked! She loves her child!*'[4] *

He hovered round his daughter. He flitted round the child: so wan, so prematurely old, *so dreadful in its gravity, so plaintive in its feeble, mournful, miserable wail.*[5] He saw the good woman tend her in the night; return to her when her grudging husband was asleep; encourage her, weep with her, set nourishment before her. He saw the day come, and the night again; the day, the night; the house of death relieved of death; the room left to herself and to the child; he heard it moan and cry; he saw it tire her out,

[1] Small capitals in *Berg*.

[2] Italic in *Berg* and in *1844*.

[3] This speech in small capitals, and the following sentence underlined, in *Berg*.

[4] *holding up his folded hands* underlined in Berg; the speech which follows is in small capitals.

[5] N.B.: from here onwards, all passages printed in *italics* are in SMALL CAPITALS in *Berg*, unless otherwise noted.

and when she slumbered, drag her back to consciousness, and hold her with its little hands upon the rack; but she was constant to it, gentle with it, patient with it. *

A change fell on the aspect of her love. One night.

She was singing faintly to the child in its sleep, and walking to and fro to hush it, when her door was softly opened, and a man looked in.

'For the last time.'

'William Fern!'

'For the last time. Margaret, my race is nearly run. I could n't finish it, without a parting word with you.* *Your child, Margaret*! Let me have it in my arms. *Let me hold your child*!'

He trembled as he took it, from head to foot.

'Is it a girl?'

'Yes.'

He put his hand before its little face.

'See how weak I'm grown, Margaret, when I want the courage to look at it! Let her be a moment. I won't hurt her. It's long ago, but—What's her name?'

'Margaret.'

He seemed to breathe more freely; and took away his hand. But he covered the child's face again, immediately.

'Margaret, it's Lilian over again.'

'Lilian!'

'I held just such another face in my arms when Lilian's mother died and left her.'

'*When Lilian's mother died and left her*!'

When he was gone, she sank into a chair, and pressed the infant to her breast. * She paced the room with it the livelong night, hushing it and soothing it. She said at intervals, '*Like Lilian, when her mother died and left her*!' And then it was that something fierce and terrible began to mingle with her love.

She dressed the child next morning with unusual care—ah vain expenditure of care upon such squalid robes!—and once more tried to find some means of life. *In vain.* *

It was a bleak, dark, cutting night: when, pressing the child close to her for warmth, she arrived outside the house she called her home. She was so faint and giddy, that she saw no one standing in the doorway until she was about to enter. Then she recognised the master of the house.

'Oh! you have come back?'

She appealed to him from the child in her arms.

'Don't you think you have lived here long enough without paying any rent?'

She repeated the same mute appeal.

'Now I see what you want, and what you mean. You know there are two parties in this house about you. But you shan't come in. That I am determined.'

She put her hair back with her hand, and looked at the sky, and the dark lowering distance. *

'*Follow her! To desperation!*'

The Bell-figures hovered in the air, and pointed where she went, down the dark street.

'She loves it! Chimes! She loves it!'

'*Follow her!*' The shadows swept upon the track she had taken, like a cloud.

He joined in the pursuit; he kept close to her; he looked into her face. He saw the fierce and terrible expression mingling with her love, and kindling in her eyes. He heard her say, '*Like Lilian! To be changed like Lilian!*' *

Putting its tiny hand up to her neck, and holding it there, within her dress: next to her distracted heart: she set its sleeping face against her, and sped onward *to the river.*

To the rolling River, swift and dim, where Winter Night sat brooding like the last dark thoughts of many who had sought a refuge there before her. Where lights upon the banks gleamed sullen, red, and dull, like torches that were burning there, to show the way to Death.

Her father followed her. She paused a moment on the river's margin. He implored the figures in the Bells now hovering above them.

'Goblins of the Bells! Spirit of the Chimes! I *have* learnt it! I *have* learnt it from the creature dearest to my heart! Oh save her, save her! * Spirits of the Chimes, think what her misery must have been, when such seed bears such fruit! Heaven meant her to be Good. Oh have mercy on my child, who, even at this pass, means mercy to her own, and dies herself, and perils her Immortal Soul, to save it!'

He could touch her, now. He could hold her, now. His strength was like a giant's.[1]

'*Spirits of the Chimes! I know that* we must trust and hope, and neither doubt ourselves, nor doubt the Good in one another. I have learnt it from the creature dearest to my heart. I clasp her in my arms again. Oh Spirits, merciful and good, I am grateful!'

He might have said more, but the Bells; the old familiar Chimes; began to ring the joy-peals for a New Year, so lustily, so merrily, so happily, so gaily, that he leaped upon his feet, and broke the spell that bound him.

'*And whatever you do, father,*' said Meg, '*don't eat tripe again, without asking some doctor whether it's likely to agree with you; for how you have*[2] *been going on, Good gracious!*'

[1] The first two sentences of this paragraph are printed in small capitals in *Berg*, but the whole paragraph is also underlined discontinuously.

[2] *have* italic in *Berg* and in *1844*; the rest of the speech in small capitals in *Berg*.

She was working with her needle, at the little table by the fire; dressing her simple gown with ribbons for her wedding. So quietly happy, so blooming and youthful, so full of beautiful promise, that he uttered a great cry as if it were An Angel in his house.

But he caught his feet in the newspaper, which had fallen on the hearth when he had fallen asleep: and somebody came rushing in between them.

'No! Not even you. Not even you. The first kiss of Meg in the New Year is mine. I have been waiting outside the house, this hour, to hear the Bells and claim it. Meg, my precious prize, a happy year! A life of happy years, my darling wife! * To-day is our Wedding Day!' *

'Richard my boy!' *cried Trotty, in an ecstacy.*[1] 'You was turned up Trumps originally; and Trumps you must be till you die!'

The child, who had been awakened by the noise, came running in half-dressed.

'Why, here she is!' *cried Trotty, catching her up.* 'Ha, ha, ha! Here's little Lilian! Here we are and here we go! Oh here we are and here we go again! And Uncle Will too! Oh, *Uncle Will, the Vision that I've had to-night, through lodging you*!'

Before Will Fern could reply, a Band of Music burst into the room, attended by the marrowbones and cleavers, the handbells, and a flock of neighbours. † They were ready for a dance in half a second, and the Drum in the Band of Music was on the brink of leathering away with all his power; when a combination of prodigious sounds was heard outside, and a comely matron came running in, attended by a man bearing a stone pitcher of terrific size.[2]

'It 's Mrs. Chickenstalker!'

'Married, and not tell me, Meg! Never! So here I am; and as it 's New Year's morning, and the morning of your wedding too, my dear, I had a little flip made, and brought it with me.'

Mrs. Chickenstalker's notion of *a little flip*, did honour to her character. The pitcher reeked like a volcano; and the man who carried it was faint.

'Mrs. Tugby! I *should*[3] say, Chickenstalker—Bless your heart and soul! A happy New Year, and many of 'em! Mrs. Tugby—I *should* say, Chickenstalker—This is William Fern and Lilian.'

'Not Lilian Fern whose mother died in Dorsetshire!'

Her uncle answered 'Yes,' and meeting, they exchanged some words; of which the upshot was, that Mrs. Chickenstalker shook him by both hands, and took the child to her capacious breast.

[1] Underlined, not in small capitals, in *Berg*; so, just below is *cried Trotty, catching her up*. The words *in an ecstacy* are not in *1844*: an unusual example of Dickens's adding, instead of deleting, such a phrase describing a character's tones.

[2] The audience roared at this entry of Mrs. Chickenstalker, and there were renewed bursts of laughter during the paragraph below about her notion of 'a little flip' (Kent, p. 174).

[3] *should . . . should* italic in *Berg* and in *1844*.

'Will Fern! Not the friend that you was hoping to find?'

'Aye! And like to prove a'most as good a friend (if that can be) as one I found.'

'*Oh! Please to play up there. Will you have the goodness!*'

To the music of the band, the bells, the marrow-bones and cleavers, all at once; and while The Chimes were yet in lusty operation out of doors; Trotty led off Mrs. Chickenstalker down the dance, and danced it in a step unknown before or since.

Had Trotty dreamed? Or are his joys and sorrows, and the actors in them, but a dream; himself a dream; the teller of this tale a dreamer, waking but now? If it be so, O Listener, dear to him in all his visions, try to bear in mind the stern realities from which these shadows come; and in your sphere endeavour to correct, improve, and soften them. So may the Rolling Year be a Happy one to You, Happy to many more whose Happiness depends on You! So may each Year be happier than the last, and not the meanest of our brethren or sisterhood debarred their rightful share, in what our Great Creator formed them to enjoy.

THE END OF THE READING

THE STORY OF LITTLE DOMBEY

THIS was the first Reading to be created from one of the novels. For the first six weeks of his career as a professional reader, Dickens had read only from the Christmas Books. Then, on 10 June 1858, he performed *The Story of Little Dombey* and, the following week, another new programme of three shorter items, one of which (*Mrs. Gamp*) was extracted from a novel. At first *Little Dombey* was a two-hour Reading like those from the Christmas Books but, by mid-October 1858, he had shortened it (like the *Carol*, the only Christmas Book to survive by then) so that *Boots at the Holly-Tree Inn* or, more often, *The Trial from 'Pickwick'* could be accommodated in the same programme. 'It is our greatest triumph everywhere,' he reported during the first provincial tour (*N*, iii. 59), and it always remained in repertoire though it was not often performed in later years. George Dolby attributed this to his always finding it so painful to read that he never did so except by particular request and under the greatest of pressure. Another explanation may be found in Kate Field's remark that, being sad, it was the least popular of all the Readings given in America; certainly only one item was performed less often on that tour, and it was given only one performance during the final Farewell series in London, in 1870.

Dickens's deciding in 1858 to make this his first big effort to extract a Reading from one of the novels is very understandable. The narrative, being confined to the life of Paul Dombey, is of manageable length—exactly the first quarter of the novel, though many irrelevant developments could of course be omitted. Paul's death, which had ended ch. 16 and the fifth instalment of *Dombey and Son*, provides an obviously effective termination of the Reading. This episode had in 1846 'thrown a whole nation into mourning' (*Life*, p. 477), and it was, with the death of Little Nell, the standard example of his pathetic powers, then very highly esteemed. Moreover, he had already discovered that the early numbers of *Dombey* read very well, for he had read them to friends in Switzerland, during their composition, with notable success.

He experienced considerable difficulty in devising this script. Two privately printed versions survive from 1858 (*Gimbel* and *Suzannet*), both extensively amended, and in 1862 or later he indeterminately worked over another version (*Berg*). The Reading originally consisted of six Chapters, all but one of which corresponds to a chapter of the novel, usually beginning and ending with the same words of the novel's chapters, thus: Chapter I from ch. 1 of the novel; Chapter II from ch. 5; Chapter III from ch. 8; Chapter IV from chs. 11 and 12; Chapter V from ch. 14; Chapter VI from ch. 16. In *Suzannet*, the prompt-copy here reprinted, Chapter II (Paul's christening) was eventually deleted and the subsequent chapters renumbered accordingly. Minor characters were progressively eliminated, including for instance Miss Tox, who had originally been quite prominent and had been much enjoyed by audiences. One character, however, suffered few cuts, and indeed his speeches had sometimes been amplified beyond what appeared in the novel: Toots. Dickens had immediately, and rightly, guessed that Toots offered splendid opportunities for his gifts as a character-actor, and Toots proved the item's great success. The description of him in the novel set a challenge to any character-actor, and Dickens was judged to have met it triumphantly: 'a voice so deep, and a manner so sheepish, that if a lamb had roared it couldn't have been so surprising' (ch. 11). Every time Toots had to say 'How do you do?'—and Dickens increased the number of occasions for it—the audience roared its delight. Kate Field's enthusiasm was typical of most critics' reactions:

> You may have loved him since childhood, . . . but until you have made Toots's acquaintance through the medium of Dickens, you have no idea of how he looks or how he talks. When Toots puts his thumb in his mouth, looks sheepish, and roars forth, 'How are you?' I feel as the man in play must feel when, for the first time, he recognises his long-lost brother with the strawberry-mark on his left arm. Dickens's Toots bears the unmistakable strawberry-mark. (pp. 70–1).

The prominence of Toots is evidence that this Reading was often very funny, though it was concerned with the sad life and early death of Paul. The pathos was strong: a recurrent textual alteration was the substitution for 'Paul' of the more emotional 'Little Dombey'. During the final chapter, audiences wept, and there was usually a hush before the ensuing applause which, wrote one reviewer, 'would have been enthusiastic but for the feeling which mellowed it and made it almost reverent'. A less favourable view is R. H. Hutton's, representative of sophisticated critics' reaction against Dickens's pathos in his later years: in the novel's presentation of Paul's death and of the water imagery surrounding it, Dickens had quite 'fondled his own conception. He used to give it even more of the same effect of high-strung sentimental melodrama, in reading or reciting it, than the written story itself contains. We well remember the mode in which he used to read, "The golden ripples on the wall . . . [etc.]". It was precisely the pathos of the Adelphi Theatre, and made the most painful impression of pathos feeding upon itself' (*Spectator*, 18 June 1870).

The Story of Little Dombey

Five Chapters[1]

CHAPTER I

Rich Mr. Dombey sat in the corner of his Wife's darkened bedchamber in the great arm-chair by the bedside, and rich Mr. Dombey's Son lay tucked up warm in a little basket, carefully placed on a low settee in front of the fire and close to it, as if his constitution were analogous to that of a muffin, *and it was essential to toast him brown while he was very new.*

Rich Mr. *Dombey* was about eight-and-forty years of age. Rich Mr. Dombey's *son*, about eight-and-forty minutes. *Mr. Dombey* was rather bald, rather red, and rather stern and pompous. *Mr. Dombey's son* was very bald, and very red, and rather crushed and spotty in his general effect, as yet.[2] *

Mr. Dombey, exulting in the long-looked-for event, the birth of a son, jingled his heavy gold watch-chain as he sat in his blue coat and bright buttons by the side of the bed, and said:

'Our house of business will once again be not only in name but in fact Dombey and Son; Dom-bey and Son! † He will be christened Paul—of course. His father's name, Mrs. Dombey, and his grandfather's! I wish his grandfather were alive this day!' And again he said, 'Dom-bey and Son.'

Those three words conveyed the one idea of Mr. Dombey's life. The earth was made for Dombey and Son to trade in, and the sun and moon were made to give them light. Common abbreviations took new meanings in his eyes, and had sole reference to them. A.D. had no concern with anno Domini, but stood for anno Dombei—and Son. *

He had been married ten years, and until this present day on which he sat jingling his gold watch-chain in the great arm-chair by the side of the bed, had had no issue.

—To speak of. There had been a girl some six years before, and she,

[1] Dickens first wrote, above the printed title, *Six Chapters*. Later he altered *Six* to *Five*.

[2] Dickens stressed the repeated *rather* . . . *rather* . . . and *very* . . . *very* . . . and the adjectives *crushed* and *spotty*: 'With Dickens, one or two adjectives answer the purpose of a whole paint-box' (Field, p. 64). The *Peterborough Advertiser*, 22 October 1859, however, wished that he would delete 'that description of the first appearance of the infant on the stage of life, and the comparison which has been so severely and, as we think, justly condemned for indelicacy and unsuitability'.

who had stolen into the chamber unobserved, was now crouching in a corner whence she could see her mother's face. But what was a girl to Dombey and Son!

Mr. Dombey's cup of satisfaction was so full, however, that he said, 'Florence, you may go and look at your pretty brother, if you like. Don't touch him!'[1]

Next moment, the *sick* lady had opened her eyes and seen the little girl; and the little girl had run towards her; and, standing on tiptoe, to hide her face in her embrace, had clung about her with a desperate affection very much at variance with her years. The lady herself seemed to faint.

'Oh Lord bless me!' said Mr. Dombey, 'I don't like the look of this. A very ill-advised and feverish proceeding having this child here. I had better ask the Doctor if he'll have the goodness to step up stairs again.'[2] † Which he did; returning with the Doctor himself, and closely followed by his sister Mrs. Chick, a lady rather past the middle age than otherwise, but dressed in a very juvenile manner, who flung her arms round his neck, and said,

'My dear Paul! This last child is quite a Dombey! He's such a perfect Dombey!'

'Well, well! I think he *is*[3] like the family. But what is this they have told me since the child was born about Fanny herself? How is Fanny?'

'My dear Paul, there's nothing whatever wrong with Fanny. Take my word, nothing whatever. There is exhaustion, certainly, but *nothing like what I underwent myself either with George or Frederick.* An effort is necessary. That's all. Ah! If dear Fanny were a Dombey!—But I dare say, although she is not a born Dombey herself, she'll make an effort; I have no doubt she'll make an effort. Knowing it to be required of her, as a duty, of course she'll make an effort.[4] † And that effort she must be encouraged, and really, if necessary, urged to make. Now my dear Paul, come close to her with me.'

The lady lay immoveable, upon her bed, clasping her little daughter to her breast. The girl clung close about her, with the same intensity as before, and never raised her head, or moved her soft cheek from her mother's face, or looked on those who stood around, or spoke, or moved, or shed a tear.

There was such a solemn stillness round the bed; and the Doctor seemed to look on the impassive form *with so much compassion and so*

[1] 'Dickens pauses and sums up "rich Mr. Dombey and Son" in a motion of the hands and that one short command, "*Don't touch him!*"' (Field, p. 65). He made a 'Warning gesture with his forefinger' (Wright).

[2] The Dr. Parker Peps episode, which followed here, is deleted in *Suzannet* by its pages being stuck together.

[3] *is* italic in *Suzannet* and in the novel.

[4] Miss Tox's entry, which followed here, is deleted in *Suzannet* by its pages being stuck together.

little hope, that Mrs. Chick was for the moment diverted from her purpose. But presently summoning courage, and what *she* called presence of mind, *she* sat down by the bedside, and said in the tone of one who endeavours to awaken a sleeper:

'Fanny! Fanny!'

There was no sound in answer but the loud ticking of Mr. Dombey's watch and the Doctor's watch, which seemed in the silence to be running a race.

'Fanny, me dear, here's Mr. Dombey come to see you. Won't you speak to him? They want to lay your little boy in bed—the baby, Fanny, you know; you have hardly seen him yet, I think; but they can't till you rouse yourself a little. Don't you think it's time you roused yourself a little? Eh?'[1] No word or sound in answer. Mr. Dombey's watch and the Doctor's watch seemed to be racing faster.

'Now, really Fanny my dear, I shall have to be quite cross with you if you don't rouse yourself. It's necessary for you to make an effort, and perhaps a very great and painful effort which you are not disposed to make; but this is a world of effort you know, Fanny, and we must never yield, when so much depends upon us. Come! Try! I must really scold you if you don't! Fanny! Only look at me. Only open your eyes to show me that you hear and understand me; will you? Good Heaven, gentlemen, what is to be done!'

The Physician, stooping down, whispered in the *little girl's* ear. Not having understood the purport of his whisper, the little creature turned her deep dark eyes towards him.

The whisper was repeated.

'Mama!'

The little voice, familiar and dearly loved, awakened some show of consciousness, even at that ebb. For a moment, the closed eye-lids trembled, and the nostril quivered, and the faintest shadow of a smile was seen.

'Mama! O dear Mama! O dear Mama!'

The Doctor gently brushed the scattered ringlets of the child, aside from the face and mouth of the mother.

Thus, clinging fast to that frail spar within her arms, the mother drifted out upon the dark and unknown sea that rolls round all the world.[2]

[1] 'How like a Dombey [Mrs. Chick] is, in her exhortation there in the chamber of death; how she places her ear close to the mother's face in expectation of a reply; how she touches her, and almost shakes her, in order that Mrs. Dombey may be roused "to make an effort"!' (Field, p. 66).

[2] 'The descriptive death of Mrs. Dombey was beautifully read, and an emotion ran through the audience as the reader closed his book at the end of Chapter I' (*Brighton Gazette*, 18 November 1858).

CHAPTER II[1]

WE must all be weaned.[2] After that sharp season in Little Dombey's life had come and gone, it began to seem as if no vigilance or care could make him a thriving boy. In his steeple-chase towards manhood, he found it very rough riding. Every tooth was a break-neck fence, and every pimple in the measles a stone-wall to him. He was down in every fit of the whooping-cough. Some bird of prey got into his throat, instead of the Thrush; and the very chickens turning ferocious—if they have anything to do with that infant malady to which they lend their name—worried him like Tiger-cats. *

He grew to be nearly five years old. A pretty little fellow; but with something wan and wistful in his small face, that gave occasion to many significant shakes of his nurse's head. She said he was *too old-fashioned.*[3]

He was childish and sportive enough at times; but he had a strange, weird, thoughtful way, at other times, of sitting brooding in his miniature arm-chair, when he looked (and talked) like one of those terrible little Beings in the Fairy tales, who, at a hundred and fifty or two hundred years of age, fantastically represent the children for whom they have been substituted. At no time did he fall into this mood so surely, as when—his little chair being carried down into his father's room—he sat there with him after dinner, by the fire. *

On one of these occasions, when they had both been perfectly quiet for a long time, and Mr. Dombey knew that the child was awake by occasionally glancing at his eye, where the bright fire was sparkling like a jewel, little Paul broke silence thus:

'Papa! what's money?'[4]

Mr. Dombey was in a difficulty; for he would have liked to give him some explanation involving the terms circulating-medium, currency, depreciation of currency, paper, bullion, rates of exchange, value of precious metals in the market, and so forth: but looking down at the little chair, *and seeing what a long way down it was*, he answered: 'Gold, and silver, and copper. Guineas, shillings, half-pence. You know what they are?'

[1] The whole of the original Chapter II (about Paul's christening) is deleted here in *Suzannet*, the pages being attached together by stamp-edging. The chapter is also deleted in *Gimbel.* It was, however, included in performances up to, at least, mid-October 1858. The beginning of the Chapter now renumbered II gave Dickens some difficulty. The present first paragraph was deleted, and a sentence (now illegible) substituted for it; later this was deleted and the original first paragraph re-inserted, with some cuts. Yet another attempt at opening this Chapter appears in *Gimbel.*

[2] This sentence was followed by another, later deleted: 'Indeed, some of us are always being weaned all our lives long; some of us never get over it.' These sentences do not occur in the novel.

[3] Wright puts inverted commas round '*too old-fashioned*'.

[4] 'Sad, high, tired voice' (Wright).

'O yes, I know what they are. I don't mean that, Papa; I mean, what's money after all?'

'What is money after all!'

'I mean, papa, what can it do?'

[1]'You'll know better bye-and-bye, my man. Money, Paul, can do anything.' *

'It isn't cruel; is it?'

'No. A good thing can't be cruel.'

[2]'As you are so rich, if money can do anything, and isn't cruel, I wonder it didn't save me my Mama. * It can't make me strong and quite well, either. * I am so tired sometimes and my bones ache so, that I don't know what to do!'

Mr. Dombey became uneasy about this odd child, and, † in consequence of his uneasiness, resolved to send him, *accompanied by his sister Florence and a Nurse*, to board with one *Mrs. Pipchin* at Brighton—an old lady who had acquired an immense reputation as 'a great manager' of children; and the secret of whose management was, to give them everything that they didn't like, and nothing that they did.

Mrs. Pipchin had also founded great fame on being a widow-lady whose husband had broken his heart in pumping water out of some Peruvian Mines. * This was a great recommendation to Mr. Dombey. For it had a rich sound. Broke his heart of the Peruvian mines, mused Mr. Dombey. Well! a very respectable way of doing it. *[3]

This celebrated Mrs. Pipchin was a marvellous ill-favoured, ill-conditioned old lady, of a stooping figure, with a mottled face, like bad marble, a hook nose, and a hard grey eye, that looked as if it might have been hammered at on an anvil. Forty years at least had elapsed since the Peruvian mines had been the death of Mr. Pipchin; but his relict still wore black bombazeen. And she was such a bitter old lady, that one was tempted to believe there had been some mistake in the application of the Peruvian Machinery, and that all her waters of gladness and milk of human kindness had been pumped out dry, instead of the Mines.[4]

[1] The preceding sentence is half-deleted and left incomplete. Wright records what Dickens read—'Rich papa patted him on the head and said:'—and what he did '(Patting table)'.

[2] 'Sigh': and later in this speech, 'Hand to forehead' (Wright).

[3] Here, and elsewhere, Dickens gave Mr. Dombey's voice 'a hard metallic ring' (Field, p. 67). 'Finger to lips' (Wright).

[4] Dickens's text becomes uncertain here. In *Suzannet*, the page which ends with this paragraph is attached by stamp-edging to one five pages further on, which begins (apart from a manuscript interpolation) with the words 'It being a part of Mrs. Pipchin's system not to encourage . . .' *T & F*, however, prints the intervening pages, so perhaps Dickens had by then (1867–8) rejected this cut. Another possibility is that this stamp-edging deletion was made after *T & F* was printed: but, if so, the sequence of the

The Castle of this Ogress was in a steep bye-street at Brighton; where the small front gardens had the unaccountable property of producing nothing but marigolds, whatever was sown in them; and where snails were constantly discovered holding on to the street doors, and other public places they were not expected to ornament, with the tenacity of cupping-glasses. There were two other very small boarders in the house, when Little Dombey (*first called so by Mrs. Pipchin*[1]) arrived. These were one Master Bitherstone from India, and a certain Miss Pankey. As to Master Bitherstone, he objected so much to the Pipchinian system, that before Little Dombey had been established in the house five minutes, he privately asked that young gentleman if he could give him any idea of the way back to Bengal. As to Miss Pankey, *she* was disabled from offering any remark, by being in solitary confinement, for the offence of having sniffed three times in the presence of visitors. * At one o'clock there was dinner, and then this young person (a mild little blue-eyed morsel of a child, who was shampooed every morning, and seemed in danger of being rubbed away altogether) was led in from captivity by the ogress herself, and instructed *that nobody who sniffed before visitors ever went to Heaven.*[2] When this great truth had been thoroughly impressed upon her, she was regaled with rice, while all the rest had cold Pork—except Mrs. Pipchin—whose constitution required warm nourishment, and who had hot mutton chops, which smelt uncommonly nice. * Also at tea, that good lady's constitution demanded hot toast while all the rest had bread and butter. *

After breakfast next morning Master Bitherstone read aloud to the rest a pedigree from Genesis (*judiciously selected by Mrs. Pipchin*), *getting over the names with the ease and clearness of a young gentleman tumbling up the treadmill.*[3] That done, Miss Pankey was borne away to be shampoo'd; and Master Bitherstone to have something else done to him with salt water, from which he always returned, very blue and dejected. Then there were Lessons. It being a part of Mrs. Pipchin's system not to encourage a child's mind to develop itself like a young flower, but to open it by force like an oyster, the moral of all her lessons was of a violent and stunning character: the hero—always a naughty boy—seldom, in the mildest catastrophe, being finished off by anything less than a lion, or a bear. † *

narration becomes very abrupt, and the above-mentioned manuscript interpolation (beginning in mid-sentence) must be ignored. There is a great deal of textual rearrangement hereabouts, with many lengthy deletions, passages written or pasted in, etc.: one passage, for instance, appears three times, having been twice deleted. The text printed below *includes* the pages formerly deleted by stamp-edging, minus the passages deleted in them.

[1] Doubly underlined.

[2] Doubly underlined.

[3] Single underlining up to *by Mrs. Pipchin*, double underlining thereafter.

At the exemplary Pipchin, Little Dombey would sit staring in his little arm-chair by the fire, for any length of time.

Once she asked him when they were alone, what he was thinking about.[1]

'You,' said Paul, without the least reserve.

'And what are you thinking about me?'

'I have been thinking you an't like my sister. There's nobody like my sister.'

'Well! there's nobody like me, either, perhaps.'

'An't there though? I am very glad there's nobody like you!'

'Upon my word, Sir! And what else are you thinking about me?'

'I am thinking how old you must be.'

'You mustn't say such things as that, young gentleman. That'll never do.'

'Why not?' *

'Never you mind, Sir. Remember the story of the little boy that was gored to death by a mad bull for asking questions.'

'If the bull was mad, how did he know that the boy had asked questions? Nobody can go and whisper secrets to a mad bull. I don't believe that story.'

'You don't believe it, Sir?'

'No.'

'Not if it should happen to have been a tame bull, you little Infidel?'[2]

As Paul had not considered the subject in that light, and had founded his conclusions on the alleged lunacy of the bull, he allowed himself to be put down for the present. But he sat turning it over in his mind, with such an obvious intention of fixing Mrs. Pipchin presently, that even that hardy old lady deemed it prudent to retreat. *

Such was life at Mrs. Pipchin's: and Mrs. Pipchin said, *and they all said, that Little Dombey* (who watched it all from his little arm-chair by the fire), *was an old, old fashioned child.*

But as Little Dombey was no stronger at the expiration of weeks of this life, than he had been on his first arrival, a little carriage was got for him, in which he could lie at his ease, *with an alphabet and other elementary works of reference*, and be wheeled down to the sea-side.

[1] Mrs. Pipchin asked this 'in the most snappish voice possible'; Paul replied 'in the gentlest childlike voice'. His 'Why not?' was spoken 'slowly and wonderingly', and Mrs. Pipchin's 'Never you mind, sir' was 'shorter and sharper than ever'. His question about the bull followed 'in a high falsetto voice and with greater deliberation than ever' (Kent, p. 180). Mrs. Pipchin became angry, at 'You don't believe it, Sir?' (Wright).

[2] '... the comedy does not get fully under way until [this] interview between Little Dombey and the exemplary Pipchin. ... Little Paul's earnestness is inexpressibly droll' (Field, p. 68). This dialogue 'awakened invariably such bursts of hearty laughter!' (Kent, p. 179).

Consistent in his odd tastes, the child set aside a ruddy-faced lad who was proposed as the drawer of this carriage and selected, instead, a weazen old crab-faced man, who was the lad's grandfather.

With this notable attendant to pull him along, and Florence always walking by his side, he went down to the margin of the ocean every day; and there he would sit or lie in his carriage for hours together: never so distressed as by the company of children—his sister Florence alone excepted, always.

'Go away, if you please,' he would say, to any child who came to bear him company.[1] 'Thank you, but I don't want you.'

Some small voice, near his ear, would ask him how he was, perhaps.

'I am very well, I thank you. But you had better go and play, if you please.'

Then he would turn his head, and watch the child away, and would say to Florence, 'We don't want any others, do we? Kiss me, Floy.'

He had even a dislike, at such times, to the company of his nurse, and was well pleased when she strolled away, as she generally did, to pick up shells *and acquaintances*. His favorite spot was quite a lonely one, far away from most loungers; and with Florence sitting by his side at work, or reading to him, or talking to him, *and the wind blowing on his face, and the water coming up among the wheels of his bed*, he wanted nothing more.

'Floy,' he said one day, 'where's India, where the friends of that boy Bitherstone—the other boy who stays with us at Mrs. Pipchin's—live?'

'Oh, it's a long, long distance off.'

'Weeks off?'

'Yes, dear. Many weeks' journey, night and day.'

[2]'If you were in India, Floy, I should—what is it that Mama did? I forget.'

'Love me?'

'No, no. Don't I love you now, Floy? What is it?—Died. If you were in India, I should die, Floy.'

She hurriedly put her work aside, and laid her head down on his pillow, caressing him. And so would she, she said, if he were there. He would be better soon.

'Oh! I am a great deal better now! I don't mean that. I mean that I should die of being so sorry and so lonely, Floy!'[3]

Another time, in the same place, he fell asleep, and slept quietly for a long time. Awakening suddenly, he listened, started up, and sat listening.

[1] '... in a soft, drawling, half-querulous voice, and with the gravest look' (Kent, p. 181).

[2] 'Repeat to dash' (Wright).

[3] In the margin here, Dickens writes *pause*. 'That momentary pause will be very well remembered by everyone who attended this Reading' (Kent, p. 182).

Florence asked him what he thought he heard.

'I want to know what it says. The sea, Floy, what is it that it keeps on saying?'

She told him that it was only the noise of the rolling waves.

'Yes, yes.[1] But I know that they are always saying something. Always the same thing. What place is over there?' He rose up, looking eagerly at the horizon.

She told him that there was another country opposite, but he said he didn't mean that; he meant farther away—farther away!

Very often afterwards, in the midst of their talk, he would break off, to try to understand what it was that the waves were always saying; and would rise up in his couch to look towards that invisible region, far away.[2]

CHAPTER III[3]

At length Mr. Dombey, one Saturday, when he came down at Brighton to see Paul, who was then six years old, resolved to make a change, and enrol him as a small student under *Doctor Blimber*. *

Whenever a young gentleman was taken in hand by Doctor Blimber, he might consider himself sure of a pretty tight squeeze. The Doctor only undertook the charge of ten young gentlemen, but he had, always ready, a supply of learning for a hundred; and it was at once the business and delight of his life to gorge the unhappy ten with it.

In fact, Doctor Blimber's establishment was a great hot-house, in which there was a forcing apparatus incessantly at work. All the boys blew before their time. Mental green-peas were produced at Christmas, and intellectual asparagus all the year round. No matter what a young gentleman was intended to bear, Doctor Blimber made him bear to pattern, somehow or other.

This was all very pleasant and ingenious, but the system of forcing was attended with its usual disadvantages. There was not the right taste about the premature productions, and they didn't keep well. Moreover, one young gentleman, with a swollen nose and an excessively large head (the oldest of the ten who had 'gone through' everything), suddenly left off blowing one day, and remained in the establishment a mere

[1] 'Slow, faint' (Wright).

[2] Strong vertical lines in both margins, against this paragraph. 'Dickens is not a reader as others are readers. He is something better. There is a death-knell in those concluding words, "far away"' (Field, p. 69). At these words, Dickens gave a 'Shake of head' (Wright).

[3] The chapter-number was altered by Dickens from IV to 3, back to 4, and finally to 3 again—presumably reflecting periods when the chapter about the christening was omitted, restored, and finally omitted again.

stalk. *And people did say that the Doctor had rather overdone it with young Toots,*[1] *and that when he began to have whiskers he left off having brains.* †

The Doctor was a portly gentleman in a suit of black, with strings at his knees, and stockings below them. He had a bald head, highly polished; a deep voice; *and a chin so very double, that it was a wonder how he ever managed to shave into the creases.* * †

His daughter, *Miss*[2] Blimber, although a slim and graceful maid, did no soft violence to the gravity of the Doctor's house. There was no light nonsense about Miss Blimber. She kept her hair short and crisp, and wore spectacles, and *she was dry and sandy with working in the graves of deceased languages.* None of your live languages for Miss Blimber. They must be dead—stone dead—and then Miss Blimber dug them up like a Ghoule.

Mrs. Blimber, her mama, was not learned herself, but she pretended to be, and that did quite as well. *She said at evening parties, that if she could have known Cicero,*[3] *she thought she could have died contented.*

As to Mr. Feeder, B.A., Dr. Blimber's assistant, he was a kind of human barrel-organ, with a little list of tunes at which he was continually working, over and over again, without any variation. †

To Dr. Blimber's Paul was taken by his father, on an appointed day.[4] The Doctor was sitting in his portentous study, with a globe at each knee, books all round him, Homer over the door, and Minerva on the mantel-shelf. 'And how do you do, Sir,' he said to Mr. Dombey, 'and how is my little friend?' When the Doctor left off, the great clock in the hall seemed (*to Paul at least*) to take him up, and to go on saying *over and over again*, 'how, is, my, lit, tle, friend, how, is, my, lit, tle, friend.' †

'Mr. Dombey,' said Doctor Blimber, 'you would wish my little friend to acquire—'

'Everything, if you please, Doctor.'

'Yes,' said the Doctor, who with his *half-shut eyes, seemed to survey Paul with the sort of interest that might attach to some choice little animal he was going to stuff.* 'Yes, exactly. Ha! We shall impart a great variety of information to our little friend, and bring him quickly forward, I dare say. I dare say. * Permit me. Allow me to present Mrs. Blimber and my daughter *Cornelia*, who will be associated with the domestic life of our young Pilgrim to Parnassus.[5] * † Who is that *at the Door*? Oh! Come in, Toots; come in. Mr. Dombey, Sir. Our head boy, Mr. Dombey.'

[1] *young Toots* doubly underlined.

[2] *Miss* (and *Mrs.* at the beginning of the next paragraph) doubly underlined.

[3] Mrs. Blimber went 'from the bottom to the top of a vocal staircase', in pronouncing this name (Field, p. 70). In the next sentence, Dickens inserted 'classical' before 'tunes' (Wright).

[4] Doubly underlined; so is *over and over again* (below).

[5] This sentence, deleted (by mistake?), is *stetted* by multiple vertical lines in both margins—a device also used elsewhere in this script but not, apparently, in other

The Doctor might have called him their head and shoulders boy, for he was at least that much taller than any of the rest. He blushed very red at finding himself among strangers, and chuckled aloud.

'An addition to our little portico, Toots; Mr. Dombey's son.'

Young Toots blushed again; and finding, from a solemn silence which prevailed, that he was expected to say something, said to Paul with surprising suddenness, 'How are you?' This, he did, in a voice so deep, and a manner so sheepish, that if a lamb had roared it couldn't have been more surprising. †

'Take him round the house, Cornelia,' said the Doctor, when Mr. Dombey was gone, 'take him round the house, Cornelia, and familiarise him with his new sphere. Go with that young lady, Dombey.'

So, Cornelia took him to the schoolroom, where there were eight young gentlemen *in various stages of mental prostration*, all very hard at work, and very grave indeed. Toots, as an old hand, had a desk to himself in one corner: and a magnificent man, of immense age, he looked, in Little Dombey's young eyes, behind it. † He now had license to pursue his own course of study; *and it was chiefly to write long letters to himself* [1]*from persons of distinction*, addressed 'P. Toots Esquire, Brighton, Sussex', and to preserve them in his desk with great care. †

Young Toots said, with heavy good nature:

'Sit down, Dombey.'

'Thank you, Sir.'

Little Dombey's endeavouring to hoist himself on to a very high window-seat, and his slipping down again, appeared to prepare Toots's mind for the reception of a discovery.

'I say, you know—you're a very small chap.'

'Yes, Sir, I'm small. Thank you, Sir.'

For, Toots had lifted him into the seat, and done it kindly too.

'Who's your tailor?' inquired Toots, after looking at him for some moments.

'It's a woman that has made my clothes as yet. My sister's dress-maker.'

'My tailor's Burgess and Co. Fash'nable. But very dear.'

Paul had wit enough to shake his head, as if he would have said it was easy to see *that*.[2]

prompt-copies (where Dickens uses the conventional directive, *Stet*). *T & F* duly prints these passages. What may have been another *Stet* sign in this script—or, more probably, was an indication that the passages indicated were, at one time, optional inclusions—is multiple horizontal lines in both margins. *T & F* does not restore any of these cuts, nor does the present edition: often, the surrounding text has been too heavily modified for the passage to be re-integrated. *Cornelia* and *at the Door* are doubly underlined.

[1] *to himself* underlined trebly.

[2] *that* italic in *Suzannet* and in the novel.

'I say! It's of no consequence, you know, but your father's regularly rich, ain't he?'

'Yes, Sir. He's Dombey and Son.'

'And which?'[1]

'And Son, Sir.'

Mr. Toots made one or two attempts to fix the firm in his mind; but not quite succeeding, said he would get Paul to mention the name again to-morrow morning, as it was rather important. *And indeed he purposed nothing less than writing himself*[2] *a private and confidential letter from Dombey and Son immediately.* †

A gong now sounding with great fury, there was a general move towards the dining-room; † where every young gentleman had a massive silver fork, and a napkin; and all the arrangements were stately and handsome. In particular, there was a butler in a blue coat and bright buttons, *who gave quite a winey flavour to the table-beer; he poured it out so superbly.*†

Tea was served in a style no less polite than the dinner; and after tea, the young gentlemen rising and bowing, withdrew to Bed. †

There were two sharers of Little Dombey's bedroom—one named Briggs—the other Tozer.[3] In the confidence of that retreat at night, Briggs said his head ached ready to split, and that he should wish himself dead if it wasn't for his mother—and a blackbird he had at home. Tozer didn't say very much, but he sighed a good deal, and told Paul to look out, for his turn would come to-morrow. After uttering those prophetic words, he undressed himself moodily, and got into bed. †

Paul had sunk into a sweet sleep, and dreamed that he was walking hand in hand with Florence through beautiful gardens, when he found that it was a dark, windy morning, with a drizzling rain: and that the gong was giving dreadful note of preparation, down in the hall.

So, he got up directly, and † proceeded softly on his journey downstairs. As he passed the door that stood ajar, a voice from within cried 'Is that Dombey?' On Paul replying, 'Yes Ma'am:' for he knew the voice to be Miss Blimber's: Miss Blimber said, 'Come in, Dombey.' And in he went.

'Now, Dombey,' said Miss Blimber. 'I'm going out for a constitutional.'[4]

Paul wondered what that was, and why she didn't send the footman out to get it in such unfavourable weather. But he made no observation

[1] Wright deletes this question, and Paul's answer, and substitutes 'Dom-a-dom (stuttering)'.

[2] *himself* doubly underlined.

[3] 'Briggs rubs his eyes, has a low snuffled voice' (Wright).

[4] '. . . as comical as . . . can be. As Miss Blimber pronounces "con-sti-tu-tion-al" it sounds like a vocal illustration of a Virginia fence. It is here, there, and everywhere. Miss Blimber peppers Dombey with it' (Field, p. 72).

on the subject: his attention being devoted to a little pile of new books, on which Miss Blimber appeared to have been recently engaged.

'These are yours, Dombey. I am going out for a constitutional; and while I am gone, that is to say in the interval between this and breakfast, Dombey, I wish you to read over what I have marked in these books, and to tell me if you quite understand what you have got to learn.' *

They comprised a little English, and a deal of Latin—names of things, declensions of articles and substantives, exercises thereon, and rules—a trifle of orthography, a glance at ancient history, a wink or two at modern ditto, a few tables, two or three weights and measures, and a little general information. When poor Little Dombey had spelt out number two, he found he had no idea of number one; fragments of which afterwards obtruded themselves into number three, which slided into number four, which grafted itself on to number two. So that it was an open question with him whether twenty Romuluses made a Remus, or hic hæc hoc was troy weight, or a verb always agreed with an ancient Briton, or three times four was Taurus a bull.[1]

'Oh, Dombey, Dombey!' said Miss Blimber, when she came back, 'this is very shocking, you know.'

Miss Blimber expressed herself with a gloomy delight, as if she had expected this result. * She divided his books into tasks on subjects A, B, C, and D, and he did very well. †

It was hard work, resuming his studies, soon after dinner; and he felt giddy and confused and drowsy and dull. But all the other young gentlemen had similar sensations, and were obliged to resume *their* studies too. The studies went round like a mighty wheel, and the young gentlemen were always stretched upon it. *

Such spirits as Little Dombey had, he soon lost, of course. But he retained all that was strange and old and thoughtful in his character; and even became more strange and old and thoughtful. He loved to be alone, and liked nothing so well as wandering about the house by himself, or sitting on the stairs listening to the great clock in the hall. He was intimate with all the paper-hanging in the house; he saw things that no one else saw in the patterns; and found out miniature tigers and lions running up the bedroom walls.

The lonely child lived on surrounded by this arabesque-work of his musing fancy, and no one understood him. Mrs. Blimber thought him 'odd,' and sometimes the servants said that Little Dombey 'moped;' but that was all.

Unless young Toots had some idea on the subject. *

He would say to Little Dombey, *fifty times a-day*, 'I say—it's of no consequence, you know—but—how are you?'

Little Dombey would answer, 'Quite well, Sir, thank you.'

'Shake hands.'

[1] This was one of 'the passages in the story which were most applauded' in Dublin (*Saunders's News-Letter*, 26 August 1858).

Which Little Dombey, of course, would immediately do. Mr. Toots generally said again, after a long interval of staring and hard breathing, 'I say—it's not of the slightest consequence, you know, but I should wish to mention it—how are you, you know?'

To which Little Dombey would again reply, 'Quite well, Sir, thank you.'

One evening a great purpose seemed to flash on Mr. Toots. He went off from his desk to look after Little Dombey, and finding him at the window of his little bedroom, blurted out all at once, as if he were afraid he should forget it: 'I say—Dombey—what do you think about?'

'*O! I think about a great many things.*'[1]

'*Do* you though?—I don't, myself.'[2]

'*I was thinking when you came in, about last night. It was a beautiful moonlight night. When I had listened to the water for a long time, I got up, and looked out at it. There was a boat over there; the sail like an arm, all silver. It went away into the distance, and what do you think it seemed to do as it moved with the waves?*'

'Pitch?'

'*It seemed to beckon—seemed to beckon me to come.*'

This was on a Friday night; it made such a prodigious impression on Mr. Toots, that he had it on his mind as long afterwards as Saturday morning.

—And so the solitary child lived on and on, surrounded by the arabesque-work of his musing fancy, and still no one understood him. * He grew fond now, of a large engraving that hung upon the staircase, where, in the centre of a group, one figure that he knew—*a figure with a light about its head—benignant, mild and merciful—stood pointing upward.*[3] * He watched *the waves and clouds at twilight with his earnest eyes, and breasted the window of his solitary room when birds flew by, as if he would have emulated them,*[4] *and soared away.*

CHAPTER IV

WHEN the Midsummer vacation approached, no indecent manifestation of joy were exhibited by the leaden-eyed young gentlemen assembled at Dr. Blimber's. Any such violent expression as 'breaking up,' would have been quite inapplicable to that polite establishment. The young gentlemen oozed away, semi-annually, to their own homes, but they never broke up. †

Mr. Feeder, B.A., however, seemed to think that he would enjoy the

[1] This, and Paul's next two speeches, have lines against them in both margins.
[2] *Do* italic in *Suzannet*. 'I don't, myself' is not in the novel.
[3] Doubly underlined.
[4] *them* doubly underlined.

holidays very much. Mr. Toots projected a life of holidays from that time forth; for, as he regularly informed Paul every day, it was his 'last half' at Doctor Blimber's, and he was going to begin to come into his property directly. *

Mrs. Blimber was by this time quite sure that Paul was *the oddest child in the world*; and the Doctor did not controvert his wife's opinion. But he said that study would do much; and he said, 'Bring him on, Cornelia! Bring him on!'

Cornelia had always brought him on as vigorously as she could; and Paul had had a hard life of it. *But, over and above the getting through his tasks, he had long had another purpose always present to him, and to which he still held fast. It was, to be a gentle, useful, quiet little fellow, always striving to secure the love and attachment of the rest; and thus he was an object of general interest; a fragile little plaything that they all liked, and whom no one would have thought of treating roughly.* †

It was darkly rumoured that even the Butler, regarding him with favor such as that stern man had never shown to mortal boy, had mingled Porter with his table beer, to make him strong. But he couldn't change his nature, and so they all agreed that Little Dombey was 'old-fashioned'.

Over and above other extensive privileges, he had free right of entry to Mr. Feeder's room, from which apartment he had twice led Mr. Toots into the open air in a state of faintness, consequent on an unsuccessful attempt to smoke a very blunt cigar: one of a bundle which that young gentleman had covertly purchased on the shingle from a most desperate smuggler, who had acknowledged, in confidence, that two hundred pounds was the price set upon his head, dead or alive, by the Custom House. †

But, Mr. Feeder's great possession was a large green jar of snuff, which Mr. Toots had brought down as a present, at the close of the last vacation; and for which he had paid a high price, as having been the genuine property of the Prince Regent. Neither Mr. Toots nor Mr. Feeder could partake of this or any other snuff, even in the most moderate degree, without being seized with convulsions of sneezing. Nevertheless it was their great delight to moisten a box-full with cold tea, stir it up on a piece of parchment with a paper-knife, and devote themselves to its consumption then and there. *In the course of which cramming of their noses, they endured surprising torments, with the constancy of martyrs; and, drinking table-beer at intervals, felt all the glories of dissipation.* *

Going into this room one evening, when the holidays were very near, Paul found Mr. Feeder filling up the blanks in some printed letters, while others were being folded and sealed by Mr. Toots. Mr. Feeder said, 'Aha, Dombey, there you are, are you? That's yours.'

'Mine, Sir?'

'Your invitation, Little Dombey.'

Paul, looking at it, found that Doctor and Mrs. Blimber requested the pleasure of Mr. P. Dombey's company at an early party on Wednesday Evening the Seventeenth Instant; and that the hour was half-past seven o'clock; and that the object was Quadrilles. He also found that the pleasure of every young gentleman's company, was requested by Doctor *and Mrs. Blimber on the same genteel occasion.*[1]

Mr. Feeder then told him, to his great joy, that his sister was invited, and that he would be expected to inform Doctor and Mrs. Blimber, in superfine small-hand, that Mr. P. Dombey would be happy to have the honour of waiting on them, in accordance with their polite invitation.

Little Dombey thanked Mr. Feeder for these hints, and pocketing his invitation, sat down on a stool by the side of Mr. Toots, as usual. *But Little Dombey's head, which had long been ailing, and was sometimes very heavy, felt so uneasy that night, that he was obliged to support it on his hand. And yet it drooped so, that by little and little it sunk on Mr. Toots's knee, and rested there, as if it had no care to be ever lifted up again.*

That was no reason why he should be deaf; but he must have been, he thought, for, by and by, he heard Mr. Feeder calling in his ear, and gently shaking him to rouse his attention. And when he raised his head, quite scared, and looked about him, he found that Doctor Blimber had come into the room; and that the window was open, and that his forehead was wet with sprinkled water; though how all this had been done without his knowledge, was very curious indeed. *

It was very kind of Mr. Toots to carry him to the top of the house so tenderly; and Paul told him that it was. But, Mr. Toots said he would do a great deal more than that, if he could; and indeed he did more as it was: for he helped Paul to undress, and helped him to bed, in the kindest manner possible, and then sat down by the bedside and chuckled very much. How he melted away, and Mr. Feeder changed into Mrs. Pipchin, Paul never thought of asking; but when he saw Mrs. Pipchin standing at the bottom of the bed, instead of Mr. Feeder, he cried out, 'Mrs. Pipchin, don't tell Florence!'

'Don't tell Florence what, my little Dombey?'

'About my not being well.'

'No, no.'

'What do you think I mean to do when I grow up, Mrs. Pipchin?'

Mrs. Pipchin couldn't guess.

'I mean to put my money all together in one Bank—never try to get any more—go away into the country with my darling Florence—have a beautiful garden, fields, and woods, and live there with her all my life!'

'Indeed, Sir?'

'Yes. That's what I mean to do, when I—' He stopped, and pondered for a moment.

[1] Pronounced 'O casion' (Wright). The underlining here seems incomplete.

Mrs. Pipchin's grey eye scanned his thoughtful face.

—'*If*[1] *I* grow up,' said Paul. *

There was a certain calm Apothecary, who attended at the establishment, and somehow *he*[2] got into the room and appeared at the bedside, was very chatty with him, and they parted excellent friends. Lying down again with his eyes shut, he heard the Apothecary say that *there was a want of vital power* (*what was that,*[3] *Paul wondered?*) *and great constitutional weakness. That there was no immediate cause for—what? Paul lost that word. And that the little fellow had a fine mind, but was an old-fashioned boy.*[4]

What old fashion could that be, Paul wondered, that was so visibly expressed in him? *

He lay in bed all that day; but got up on the next, and went down stairs. Lo and behold, there was something the matter with the great clock; and a workman on a pair of steps had taken its face off, and was poking instruments into the works by the light of a candle! This was a great event for Paul, who sat down on the bottom stair, and watched the operation. As the workman said, when he observed Paul, 'How do you do, Sir?'[5] Paul got into conversation with him. Finding that his new acquaintance was not very well informed on the subject of the Curfew Bell of ancient days, Paul gave him an account of that institution; and also asked him, as a practical man, what he thought about King Alfred's idea of measuring time by the burning of candles? To which the workman replied, that he thought it would be the ruin of the clock trade if it was to come up again. At last the workman put away his tools, and went away. *Though not before he had whispered something, on the doormat, to the footman, in which there was the phrase 'old-fashioned'—for Paul heard it.*

[6] *What could that old fashion be, that seemed to make the people sorry! What could it be!* * †

And now it was that he began to think it must surely be old-fashioned, to be very thin and light, and easily tired, and soon disposed to lie down anywhere and rest; for he couldn't help feeling that these were more and more his habits every day.

At last the party-day arrived; and Doctor Blimber said at breakfast, 'Gentlemen, we will resume our studies on the twenty-fifth of next month.' Mr. Toots immediately threw off his allegiance, and put on a

[1] *If* trebly underlined.

[2] *he* italic in *Suzannet* and in the novel.

[3] Wright puts inverted commas round *What was that?* (and round *what?* in the next sentence).

[4] *but was an old-fashioned boy* doubly underlined.

[5] 'Loud' (Wright).

[6] This paragraph, and the next, have lines against them in both margins.

ring: and mentioning the Doctor in casual conversation shortly afterwards, spoke of him as 'Blimber'! This act of freedom inspired the older pupils with admiration; but the younger spirits seemed to marvel that no beam fell down and crushed him.

Not the least allusion was made to the ceremonies of the evening, either at breakfast or at dinner; but there was a bustle in the house all day, and Paul made acquaintance with various strange benches and candlesticks, and met a harp in a green great-coat standing on the landing outside the drawing-room door. There was something queer, too, about Mrs. Blimber's head at dinner-time, as if she had screwed her hair up too tight;[1] and though Miss Blimber showed a graceful bunch of plaited hair on each temple, she seemed to have her own little curls in paper underneath, and in a playbill too; for Paul read 'Theatre Royal' over one of her sparkling spectacles, and 'Brighton' over the other.

There was a grand array of white waistcoats and cravats in the young gentlemen's bedrooms as evening approached, and such a smell of singed hair, that Doctor Blimber sent up the footman with his compliments, and wished to know if the house was on fire. But, it was only the hairdresser curling the young gentlemen, and over-heating his tongs in the ardour of business.

When Paul was dressed he went down into the drawing-room; where he found Doctor Blimber pacing up and down full dressed, but with a dignified and unconcerned demeanour, as if he thought it barely possible that one or two people might drop in by and bye. Shortly afterwards, Mrs. Blimber appeared, looking lovely, Paul thought, and attired in such a number of skirts that it was quite an excursion to walk round her. Miss Blimber came down soon after her mama; a little squeezed in appearance, but very charming.

Mr. Toots and Mr. Feeder were the next arrivals. Each of these gentlemen brought his hat in his hand, as if he lived somewhere else; and when they were announced by the butler, Doctor Blimber said, 'Aye, aye, aye! God bless my soul!' and seemed extremely surprised to see them. Mr. Toots was one blaze of jewellery and buttons; and he felt the circumstance so strongly, that when he had shaken hands with the Doctor, and had bowed to Mrs. Blimber and Miss Blimber, he took Paul aside, and said, 'What do you think of this, Dombey!'

But notwithstanding his modest confidence in himself, Mr. Toots appeared to be involved in a good deal of uncertainty whether, on the whole, it was judicious to button the bottom button of his waistcoat; and whether, on a calm revison of all the circumstances, it was best to wear his wristbands turned up or turned down. Observing that Mr. Feeder's were turned up, Mr. Toots turned his up; but the wristbands

[1] 'Action—side of head' (Wright).

of the next arrival being turned down, Mr. Toots turned his down. The differences in point of waistcoat-buttoning, not only at the bottom, but at the top too, became so numerous and complicated as the arrivals thickened, that Mr. Toots was continually fingering that article of dress, *as if he were performing on some*[1] *instrument.*

All the young gentlemen, tightly cravatted, curled, and pumped, and with their best hats in their hands, having been at different times announced and introduced, Mr. Baps, the dancing-master, came, accompanied by Mrs. Baps, to whom Mrs. Blimber was extremely kind and condescending. Mr. Baps was a very grave gentleman; and before he had stood under the lamp five minutes, he began to talk to Toots (*who had been silently comparing pumps with him*) about what you were to do with your raw materials when they came into your ports in return for your drain of gold.[2] Mr. Toots, to whom the question seemed perplexing, suggested 'Cook 'em.' But Mr. Baps did not appear to think that would do.

Paul now slipped away from the cushioned corner of a sofa, which had been his post of observation, and went down stairs into the tea room to be ready for Florence. Presently she came: looking so beautiful in her simple ball dress, with her fresh flowers in her hand, that when she knelt down on the ground to take Paul round the neck and kiss him, he could hardly make up his mind to let her go again.

[3]'*But what is the matter, Floy?*' *asked Paul, almost sure that he saw a tear on her face.*

'*Nothing, darling; nothing.*'

Paul touched her cheek gently with his finger—and it was[4] *a tear!* '*Why, Floy!*'

'*We'll go home together, and I'll nurse you, love.*'

'*Nurse me! Floy. Do* you *think I have grown old-fashioned?*[5] *Because, I know they say so, and I want to know what they mean, Floy.*' *

From his nest among the sofa-pillows, where she came at the end of every dance, he could see and hear almost everything that passed at the Ball. * There was one thing in particular that he observed. Mr. Feeder, after imbibing several custard-cups of negus, began to enjoy himself, and told Mr. Toots that he was going to throw a little spirit into the thing.[6] After that, he not only began to dance as if he meant dancing and nothing else, but secretly to stimulate the music to perform

[1] 'some *curious* instrument' (Wright).

[2] Wright puts inverted commas round Mr. Baps's enquiry ('what you were . . . drain of gold').

[3] Lines in both margins, against this dialogue.

[4] *was* italic in *Suzannet* and the novel. Wright substitutes 'the child' for 'Paul' in this sentence.

[5] This sentence doubly underlined; *you* italic in *Suzannet* and in the novel.

[6] Wright puts inverted commas round 'going to throw a little spirit into the thing'.

wild tunes. Further, he became particular in his attentions to the ladies; and dancing with Miss Blimber, whispered to her—whispered to her!—though not so softly but that Paul heard him say this remarkable poetry,

> 'Had I a heart for falsehood framed,
> I ne'er could injure You!'

This, Paul heard him repeat to four young ladies, in succession. Well might Mr. Feeder say to Mr. Toots, that he was afraid he should be the worse for it tomorrow! * †

A buzz at last went round of 'Dombey's going!' 'Little Dombey's going!' and there was a general move after him and Florence down the staircase and into the hall. * Once for a last look he turned, surprised to see how shining, and how bright and numerous the faces were, and how they seemed like *a great dream, full of eyes*. †

There was much, soon afterwards—next day, and after that—which Paul could only recollect confusedly. As, why they stayed at Mrs. Pipchin's days and nights, instead of going home.

But he could remember, when he got to his old London home and was carried up the stairs, that there had been the rumbling of a coach for many hours together, while he lay upon the seat, with Florence still beside him, and old Mrs. Pipchin sitting opposite. He remembered his old bed too, when they laid him down in it. But, there was something else, and recent, that still perplexed him.

'I want to speak to Florence, if you please. To Florence by herself, for a moment!'

She bent down over him, and the others stood away.

'Floy, my pet, wasn't that Papa in the hall, when they brought me from the coach?'

'Yes, dear.'

'He didn't cry, and go into his room, Floy, did he, when he saw me coming in?'

She shook her head, and pressed her lips against his cheek.

'I'm very glad he didn't cry, Floy. I thought he did. Don't tell them that I asked.'

CHAPTER THE LAST[1]

LITTLE Dombey had never risen from his little bed. He lay there, listening to the noises in the street, quite tranquilly; not caring much how the time went, but watching it and watching everything.

When the sunbeams struck into his room through the rustling blinds,

[1] Dickens altered VI to V, and then wrote 'The Last'. According to Wright, he always, 'in other readings' too, preluded his final section by saying: 'Chapter the Last'.

and quivered on the opposite wall, like golden water, he knew that evening was coming on, and that the sky was red and beautiful. As the reflection died away, and a gloom went creeping up the wall, he watched it deepen, deepen, deepen, into night. Then, he thought how the long unseen streets were dotted with lamps, and how the peaceful stars were shining overhead. *His fancy had a strange tendency to wander to the River, which he knew was flowing through the great city; and now he thought how black it was, and how deep it would look, reflecting the hosts of stars—and more than all, how steadily it rolled away to meet the sea.*[1]

As it grew later in the night, and footsteps in the street became so rare that he could hear them coming, count them as they passed, and lose them in the hollow distance, he would lie and watch the many-coloured ring about the candle, and wait patiently for day. *His only trouble was, the swift and rapid river.*[2] *He felt forced, sometimes, to try to stop it—to stem it with his childish hands—or choke its way with sand—and when he saw it coming on, resistless, he cried out!*[3] But a word from Florence, who was always at his side, restored him to himself; and leaning his poor head upon her breast, he told Floy of his dream, and smiled.

When day began to dawn again, he watched for the sun; and when its cheerful light began to sparkle in the room, he pictured to himself—pictured! he saw—the high church towers rising up into the morning sky, the town reviving, waking, starting into life once more, the river glistening as it rolled (*but rolling fast as ever*), and the country bright with dew. Familiar sounds and cries came by degrees into the street below; the servants in the house were roused and busy; faces looked in at the door, and voices asked his attendants softly how he was. Paul always answered for himself, '*I am better. I am a great deal better, thank you! Tell Papa so!*'

By little and little, he got tired of the bustle of the day, the noise of carriages and carts, and people passing and re-passing: and would fall asleep, *or be troubled with a restless and uneasy sense again.* '*Why, will it never stop, Floy?*' he would sometimes ask her. '*It is bearing me away, I think!*'

But, she could always soothe and reassure him; and it was his daily delight to make her lay her head down on his pillow, and take some rest.

'You are always watching me, Floy. Let *me* watch *you*,[4] now!'

They would prop him up with cushions in a corner of his bed, and

[1] Lines in both margins, against this sentence.

[2] 'Quick' (Wright). Lines in both margins, against these two underlined sentences.

[3] *he cried out* underlined trebly. This 'one startling instant' apart, all the pathetic incidents were narrated *sotto voce*, writes Kent (p. 185); but, on these three words, Dickens's voice was suddenly raised in 'an abrupt outcry'. 'Dropping his voice . . . instantly afterwards, to the gentlest tones, . . . the Reader continued in those subdued and tender accents to the end'.

[4] *me* underlined; *you* italic in *Suzannet* and in the novel.

there he would recline, the while she lay beside him: bending forward often-times to kiss her, and whispering to those who were near that she was tired, and how she had sat up so many nights beside him.

Thus, the flush of the day, in its heat and light, would gradually decline; and again the golden water would be dancing on the wall. †

The people around him changed unaccountably—and what had been the Doctor, would be his father, sitting with his head leaning on his hand. *This figure with its head leaning on its hand returned so often, and remained so long, and sat so still and solemn, never speaking, never being spoken to, and rarely lifting up its face, that Paul began to wonder languidly, if it were real.*

'Floy! What *is*[1] that?'

'Where, dearest?'

'There! at the bottom of the bed.'

'There's nothing there, except Papa!'

The figure lifted up its head, and rose, and coming to the bedside, said:

'My own boy! Don't you know me?'[2]

Paul looked it in the face. Before he could reach out both his hands to take it between them and draw it towards him, the figure turned away quickly from the little bed, and went out at the door.

The next time he observed the figure sitting at the bottom of the bed, he called to it.

'Don't be so sorry for me, dear Papa! Indeed I am quite happy!'

His father coming, and bending down to him, he held him round the neck, and repeated those words to him several times, and very earnestly; and he never saw his father in his room again at any time, whether it were day or night, but he called out, 'Don't be so sorry for me! Indeed I am quite happy!' This was the beginning of his always saying in the morning that he was a great deal better, and that they were to tell his father so.

How many times the golden water danced upon the wall; how many nights the dark river rolled towards the sea in spite of him; Paul never sought to know. If their kindness or his sense of it, could have increased, they were more kind, and he more grateful every day; *but, whether there were many days or few, appeared of little moment now, to the gentle boy.*

One night he had been thinking of his mother and her picture in the drawing-room down stairs. The train of thought suggested to him to inquire if he had ever seen his mother? For, *he could not remember whether they had told him yes, or no; the river running very fast, and confusing his mind.*

'Floy, did I ever see mama?'

'No, darling, why?'

[1] *is* italic in *Suzannet* and in the novel. [2] 'Despairingly' (Wright).

'Did I never see any kind face, like a mama's, looking at me when I was a baby, Floy?'

'Oh yes, dear!'

'Whose, Floy?'

'Your old nurse's. Often.'

'And where is my old nurse? Show me that old nurse, Floy, if you please!'

'She is not here, darling. She shall come tomorrow.'

'Thank you, Floy!'

Little Dombey closed his eyes with those words, and fell asleep. When he awoke, the sun was high, and the broad day was clear and warm. Then he awoke[1]—woke mind and body—and sat upright in his bed. He saw them now about him. There was no gray mist before them, as there had been sometimes in the night. He knew them every one, and called them by their names.

'And who is this? Is this my old nurse?' asked the child, regarding with a radiant smile, a figure coming in.

Yes, yes. No other stranger would have shed those tears at sight of him, and called him her dear boy, her pretty boy, her own poor blighted child.[2] No other woman would have stooped down by his bed, and taken up his wasted hand, and put it to her lips and breast, as one who had some right to fondle it. No other woman would have so forgotten everybody there but him and Floy, and been so full of tenderness and pity.

'Floy! this is a kind good face! I am glad to see it again. Don't go away, old nurse! Stay here! * Good-bye!'

'Good-bye, my child?' cried Mrs. Pipchin, hurrying to his bed's head. 'Not good-bye?'

'Ah, yes! Good-bye!—Where is Papa?'

His father's breath was on his cheek, before the words had parted from his lips. The feeble hand waved in the air, as if it cried, 'good-bye!' again.

'Now lay me down; and Floy, come close to me, and let me see you!'

Sister and brother wound their arms around each other, and the golden light came streaming in, and fell upon them, locked together.[3]

'How fast the river runs, between its green banks and the rushes, Floy! But, it's very near the sea now. I hear the waves! They always said so!'

Presently, he told her that the motion of the boat upon the stream was

[1] 'Then he awoke' is inserted in manuscript, to cover two short deletions. Dickens nodded here; he forgot that Paul was already awake, and had dropped off to sleep again only in the deleted passage.

[2] 'Feelingly'; at 'taken up his wasted hand ...', Dickens put 'Hand to lips', 'Hand to breast' (Wright).

[3] This paragraph is underlined in *Gimbel.*

lulling him to rest. Now, the boat was out at sea. *And now, there was a shore before him. Who stood on the bank!—*[1]

He put his hands together, as he had been used to do, at his prayers. He did not remove his arms to do it; but they saw him fold them so, behind his sister's neck.

'Mama is like you, Floy. I know her by the face! But tell them that the picture on the stairs at school, is not Divine enough.[2] The light about the head is shining on me as I go!'

The golden ripple on the wall came back again, and nothing else stirred in the room. The old, old, fashion! The fashion that came in with our first garments, and will last unchanged until our race has run its course, and the wide firmament is rolled up like a scroll. The old, old fashion—Death![3]

Oh, thank GOD, all who see it, for that older fashion yet, of Immortality! And look upon us, Angels of young children, with regards not quite estranged, when the swift river bears us to the ocean!

[1] The whole of this paragraph is underlined in *Gimbel.*

[2] Wright underlines *Divine enough*, but comments: 'Too loud altogether'.

[3] *Death* underlined five times—but spoken 'In a low whispering voice' (Wright). In *Gimbel*, too, the whole paragraph is underlined, with special emphasis on *Death.* The ensuing, and final, paragraph (all underlined in *Gimbel*) was spoken 'in a tearful voice' (Kent, p. 187). See the headnote to this Reading for R. H. Hutton's severe criticism of Dickens's performance at this point ('Adelphi Theatre' pathos).

THE POOR TRAVELLER

ON 17 June 1858 Dickens for the first time departed from his practice of performing only one item, and instead read three short pieces: *The Poor Traveller*, *Boots at the Holly-Tree Inn* and *Mrs. Gamp*, the prompt-copy for all of which is in *Berg*. Abbreviated from the first tale in the *Household Words* Christmas number for 1854, *The Poor Traveller* is a feeble and uncharacteristic piece, though its pathos and melodrama made some appeal to the taste of its time: 'one of the most beautiful little stories that ever was written, either by Mr. Dickens or anybody else. We do not envy the man who can read it, and not feel a tear start unbidden to his eye' (*Bradford Observer*, 21 October 1858). Dickens had given a successful pre-publication reading of it to friends, in 1854, and maybe memories of this occasion inclined him to choose it in 1858: but probably he also wanted a contrasting item in this triple bill. *Boots* and *Mrs. Gamp* were predominantly comic; *The Poor Traveller*, different in mode and tone, enabled him to exercise elocutionary gifts which the rest of his current repertoire did not demand.

Some audiences found his pathos here 'deep and thrilling', and they 'frequently interrupted by applause' the more stirring scenes in his narration. Other critics judged that, on this evidence, Dickens's forte as a reader did not lie in the pathetic and that his more melodramatic passages, though effective, depended upon 'tricks of a rather puerile kind'. The performance, making heavy use of emphatic references to Captain Taunton's 'deep dark eyes', 'perpetually bringing in the name "Richard Doubledick"', and other such devices, reminded the *Saturday Review* (19 June 1858) of what it saw as Dickens's stylistic weaknesses—'the numberless instances in which he has, as we think, thrown away the genuine success he might have achieved, by having recourse to the paltry artifices of stage effect'. After his first season (1858–9), this item virtually disappeared from Dickens's repertoire.

The Poor Traveller

IN the year one thousand seven hundred and ninety-nine,[1] a relative of mine went limping down, on foot, to the town of Chatham. He was a poor traveller, with not a farthing in his pocket, and he slept that night at an old foundation for Poor Travellers, where some Poor Travellers sleep every night.

He went down to Chatham, to enlist in a cavalry regiment, if a cavalry regiment would have him; if not, to take King George's shilling from any corporal or sergeant who would put a bunch of ribbons in his hat. *His object was, to get shot; but, he thought he might as well ride to death as be at the trouble of walking.*[2]

His Christian name was Richard, but he was better known as Dick. He dropped his own surname on the road down, and took up that of Doubledick. He passed as Richard Doubledick; age twenty-two; height, five foot ten. There was no cavalry in Chatham, so he enlisted into a regiment of the line, *and was glad to get drunk and forget all about it.*

He had gone wrong and run wild. His heart was in the right place, but it was sealed up. He had been betrothed to a good and beautiful girl whom he had loved better than she—or perhaps even he—believed; but, in an evil hour, he had given her cause to say to him, 'Richard, I will never marry any other man. I will live single for your sake, but Mary Marshall's lips;' her name was Mary Marshall; 'never address another word to you on earth. Go, Richard! Heaven forgive you!' This finished him. This took him down to Chatham. This made him Private[3] Doubledick, with a determination to be shot.

There was not a more reckless soldier in Chatham barracks, in the year one thousand seven hundred and ninety-nine, than Private Doubledick. He associated with the dregs of every regiment, he was as seldom sober as he could be, and was constantly under punishment. It became clear to the whole barracks, that Private Doubledick would very soon be flogged.

[1] Here, and at the beginning of the fifth paragraph, Dickens altered the date to 1779; in the second instance, he altered it back again to 1799, but he forgot to do so here.

[2] All the underlining in this prompt-copy (unless otherwise noted) is in Dickens's double interrupted style, and will be represented here throughout by italic without any footnoting.

[3] The deletion of 'Richard' throughout presumably took place after the initial performance reviewed in the *Saturday Review*, in which the emphatic repetition of 'Richard Doubledick' was noted (quoted above, in the headnote).

Now, *the Captain* of Doubledick's company was a young gentleman not above five years his senior, whose eyes had an expression in them which affected Private Doubledick in a very remarkable way.[1] They were bright, handsome, dark eyes—what are called laughing eyes generally, and, when serious, rather steady than severe—but, they were the only eyes now left in his narrowed world that Private Doubledick could not stand. He could not so much as salute *Captain Taunton* in the street, like any other officer. In his worst moments he would rather turn back and go any distance out of his way, than encounter those two handsome, dark, bright eyes.

One day, when Private Doubledick came out of the Black hole—where he had been passing the last eight-and-forty hours—and in which retreat he spent a good deal of his time—he was ordered to betake himself to Captain Taunton's quarters. In the squalid state of a man just out of the Black hole, he had less fancy than ever for being seen by the Captain; but, he was not so mad yet as to disobey orders, and consequently went up to the officers' quarters: twisting and breaking in his hands as he went along, a bit of straw that had formed the decorative furniture of the Black hole.

'Come in!' Private Doubledick pulled off his cap, took a stride forward, and stood in the light of the dark bright eyes.

'Doubledick, do you know where you are going to?'

'To the Devil, Sir?'

'Yes. And very fast.'

He turned the straw of the Black hole in his mouth, and made a miserable salute of acquiescence.[2]

'Doubledick, since I entered his Majesty's service, a boy of seventeen, I have been pained to see many men of promise going that road; but, I have never been *so* pained to see a man determined to make the shameful journey, as I have been, ever since *you* joined the regiment, to see *you*.'[3]

'I am only a common soldier, sir. It signifies very little what such a poor brute comes to.'

[1] '... when the Reader came ... to these words ... the effect was singularly striking. Out of the Reader's own eyes would look the eyes of that Captain, as the Author himself describes them' (Kent, p. 197). Many reminiscences of Dickens, when met socially or seen on a platform, recall the extraordinary intensity of his eyes; *The Poor Traveller* was the only Reading which gave him such a ready-made opportunity to exploit this feature of his physical presence.

[2] '... one did not hear the words simply, one saw it done precisely as it is described'. This interview with the Captain was 'one of the most exquisitely portrayed incidents in the whole of this Reading' (Kent, pp. 198–9).

[3] Here Dickens deleted, in *Berg*, this paragraph: 'Private Richard Doubledick began to find a film stealing over the floor at which he looked; also to find the legs of the Captain's breakfast-table turning crooked, as if he saw them through water.' As Kent remarked, Dickens obtained 'pretty nearly the effect' of these words, by the way he spoke—after a momentary pause—the speech that follows.

'You are a man of education; and if you say that, meaning what you say, you have sunk lower than I had believed. How low that must be, I leave you to consider: knowing what I know of your disgrace, and seeing what I see.'

'I hope to get shot soon, sir, and then the regiment, and the world together, will be rid of me.'

Looking up he met the eyes that had so strong an influence over him. He put his hand before his own eyes, and the breast of his disgrace-jacket swelled as if it would fly asunder.

'I would rather see this in you, Doubledick, than I would see five thousand guineas counted out upon this table for a gift to my good mother. Have you a mother?'

'I am thankful to say she is dead, sir.'

'If your praises were sounded from mouth to mouth through the whole regiment, through the whole army, through the whole country, you would wish she had lived, to say with pride and joy, "He is my son!"'

'Spare me, sir. She would never have heard any good of me. She would never have had any pride and joy in owning herself my mother. Love and compassion she might have had, and would have always had, I know; but not——Spare me, sir! I am a broken wretch, quite at your mercy!'[1]

'My friend——'

'God bless you, sir!'

'You are at the crisis of your fate, my friend. Hold your course unchanged a little longer, and you know what must happen. *I*[2] know even better than you can imagine, that after that has happened, you are lost. No man who could shed such tears, could bear such marks.'

'I fully believe it, sir.'

'But a man in any station can do his duty, and, in doing it, can earn his own respect, even if his case should be so very unfortunate and so very rare, that he can earn no other man's. A common soldier—poor brute though you called him just now—has this advantage in the stormy times we live in, that he always does his duty before a host of sympathising witnesses. Do you doubt that he may so do it as to be extolled through a whole regiment, through a whole army, through a whole country? Turn while you may yet retrieve the past, and try.'

'I will! I ask for only one witness, sir.'

[1] Again action replaced speech here. Dickens deleted in *Berg* the next sentence: 'And he turned his face to the wall and stretched out his imploring hand.' Kent remarks: 'How eloquently that "imploring hand" spoke in the agonised, dumb supplication of its movement, coupled as it was with the shaken frame and the averted countenance, those who witnessed this Reading will readily recall to their recollection' (p. 200).

[2] *I* italic in *Household Words* (*HW*) and in *Berg*.

'I understand you. I will be a watchful and a faithful one.'

I have heard from Private Doubledick's own lips, that he dropped down upon his knee, kissed that officer's hand, arose, *and went out of the light of the dark bright eyes, an altered man.*[1]

In that year, one thousand seven hundred and ninety-nine, the French were in Egypt, in Italy, in Germany, where not? Napoleon Buonaparte had likewise begun to stir against us in India. In the very next year, when we formed an alliance with Austria against him, Captain Taunton's regiment was on service in India. And there was not a finer non-commissioned officer in it—no, nor in the whole line—than *Corporal* Doubledick.

In eighteen hundred and one, the Indian army were on the coast of Egypt. Next year was the year of the proclamation of the short peace, and they were recalled. It had then become well-known to thousands of men, that wherever Captain Taunton led, there, close to him, ever at his side, firm as a rock, true as the sun, and brave as Mars, would be certain to be found, while life beat in their hearts, that famous soldier, *Sergeant* Doubledick.

Eighteen hundred and five saw wonders done by a *Sergeant-Major*, who cut his way single-handed through a solid mass of men, recovered the colours of his regiment, and rescued his wounded captain, who was down in a jungle of horses' hoofs and sabres—that year saw such wonders done, I say, by this brave Sergeant-Major, that he was specially made the bearer of the colours he had won; and *Ensign* Doubledick had risen from the ranks.

Sorely cut up in every battle, but always reinforced by the bravest of men, this regiment fought its way through the Peninsular war, up to the investment of Badajos in eighteen hundred and twelve. † One day,—not in the great storming, but in repelling a hot sally of the besieged upon our men at work in the trenches, who had given way,—the two officers, Major Taunton and Ensign Doubledick, found themselves hurrying forward, face to face, against a party of French infantry who made a stand. There was an officer at their head, encouraging his men,—a courageous, handsome, gallant officer of five and thirty—whom Doubledick saw hurriedly, but saw well. He particularly noticed this officer waving his sword, and rallying his men with an eager and excited cry, when they fired, *and Major Taunton dropped.*

It was over in ten minutes more, and Doubledick returned to the spot where he had laid the best friend man ever had, on a coat spread upon the wet clay. Major Taunton's uniform was opened at the breast, and on his shirt were three little spots of blood.

'Dear Doubledick, I am dying.'

'For the love of Heaven, no! Taunton! My preserver, my guardian

[1] Spoken 'in terms how manly and yet how tender in their vibration' (Kent, p. 202).

angel, my witness! Dearest, truest, kindest of human beings! Taunton! For God's sake!'[1]

The bright dark eyes—so very, very dark now, in the pale face—smiled upon him; *and the hand he had kissed thirteen years ago, laid itself on his breast.*

'Write to my mother. You will see Home again. Tell her how we became friends. It will comfort her, as it comforts me.'

He spoke no more, but faintly signed for a moment towards his hair as it fluttered in the wind. The Ensign understood him. He smiled again when he saw that, and gently turning his face over on the supporting arm as if for rest, *died, with his hand upon the breast in which he had revived a soul.*

No dry eye looked on Ensign Doubledick, that melancholy day. He buried his friend on the field, and became a lone man. Beyond his duty he appeared to have but two remaining cares in life; one, to preserve the little packet of hair he was to give to Taunton's mother; *the other, to encounter that French officer who had rallied the men under whose fire Taunton fell.* A new legend now began to circulate among our troops; and it was, that *when he and the French officer came face to face once more, there would be weeping in France.*

The war went on until the Battle of Toulouse was fought. In the returns sent home, appeared these words: 'Severely wounded, but not dangerously, *Lieutenant* Doubledick.'

At Midsummer time in the year eighteen hundred and fourteen, Lieutenant Doubledick, now a browned soldier, seven and thirty years of age, came home to England, invalided. He brought the hair with him, near his heart. *Many a French officer had he seen, since that day; many a dreadful night, in searching with men and lanterns for his wounded, had he relieved French officers lying disabled; but, the mental picture and the reality had never come together.*

Though he was weak and suffered pain, he lost not an hour in getting down to Frome in Somerset-shire, where Taunton's mother lived. *In the sweet, compassionate words of the most compassionate of books, 'he was the only son of his mother, and she was a widow.'* †

Never, from the hour when he enlisted at Chatham, had he breathed his right name, or the name of Mary Marshall, or a word of the story of his life, into any ear, except his reclaimer's. But, that night, remembering the words he had cherished for two years, 'Tell her how we became friends. It will comfort her, as it comforts me,' he related everything. It gradually seemed to him, as if in his maturity he had recovered a mother; it gradually seemed to her, as if in her bereavement she had found

[1] 'To listen to that agonised entreaty as it started from the trembling and one could almost have fancied whitened lips of the Reader, was to be with him there upon the instant on the far-off battle-field' (Kent, p. 203).

a son. When he was able to rejoin his regiment in the spring, he left her; thinking was this indeed the first time he had ever turned his face towards the old colours with a woman's blessing!

He followed the old colours—so ragged, scarred and pierced with shot, that they would scarcely hold together—to Quatre Bras, and to Ligny. He stood beside them, in an awful stillness of many men, shadowy through the mist and drizzle of a wet June forenoon, on the field of Waterloo.

The famous regiment was in action early in the battle, and received its first check in many an eventful year, when *he* was seen to fall. But, it swept on to avenge him, and left behind it no such creature in the world of consciousness, as Lieutenant Doubledick.

Through pits of mire and pools of rain; along deep ditches, once roads, that were ploughed to pieces by artillery, heavy waggons, tramp of men and horses, and the struggle of every wheeled thing that could carry wounded soldiers; jolted among the dying and the dead, so disfigured by blood and mud as to be hardly recognizable for humanity; undisturbed by the moaning of men and the shrieking of horses, which, newly taken from the peaceful pursuits of life, could not endure the sight of the stragglers lying by the wayside, never to resume their toilsome journey; dead, as to any sentient life that was in it, and yet alive; the form that had been Lieutenant Doubledick, with whose praises England rang, was conveyed to Brussels. There, it was tenderly laid down in hospital: and there it lay, week after week, through the long bright summer days, until the harvest had ripened and was gathered in. †

Slowly labouring, at last, through a long heavy dream of confused time and place, Lieutenant Doubledick came back to life. To the beautiful life of a calm autumn-evening sunset. To the peaceful life of a fresh quiet room with a large window standing open; a balcony beyond, in which were moving leaves and sweet-smelling flowers; beyond again, the clear sky, and the sun, pouring its golden radiance on his bed.

It was so tranquil and so lovely, that he thought he had passed into another world. And he said in a faint voice, 'Taunton, are you near me?'

A face bent over him. Not his; his mother's.

'I came to nurse you. We have nursed you, many weeks. You were moved here, long ago. Do you remember nothing?'

'Nothing. Where is the regiment? What has happened?'

'A great victory, dear. The war is over, and the regiment was the bravest in the field.'

'Thank God! Was it dark just now?'

'No.'

'It was only dark to me? Something passed away, like a black shadow. As it went, and the sun touched my face, I thought I saw a light white cloud pass out at the door. Was there nothing that went out?'

She shook her head, and, in a little while, he fell asleep, and from that time, he recovered. Slowly, for he had been desperately wounded in the head, and had been shot in the body.

One day, he awoke out of a sleep, refreshed, and asked Mrs. Taunton to read to him. But, the curtain of the bed softening the light was held undrawn; and a woman's voice spoke, which was not hers.

'Can you bear to see a stranger? Will you like to see a stranger?'[1]

'Stranger!' *The voice awoke old memories, before the days of Private Doubledick.*

'A stranger now, but not a stranger once. Richard, dear Richard, lost through so many years, my name——'

He cried out her name, 'Mary!' and his head lay on her bosom.

'I am not breaking a rash vow, Richard. These are not Mary Marshall's lips that speak. I have another name.'

She was married.

'I have another name, Richard. Did you ever hear it?'

'Never!'

'Think again, Richard. Are you sure you never heard my altered name?'

'Never!'

'Don't move your head to look at me, dear Richard. Let it lie here, while I tell my story. I loved a generous, noble man with my whole heart; loved him for years and years, faithfully, devotedly; with no hope of return; knowing nothing of his highest qualities—not even knowing that he was alive. He was a brave soldier. He was honoured and beloved by thousands, when the mother of his dear friend found me, and showed me that in all his triumphs he had never forgotten me. He was wounded in a great battle. He was brought, dying, here into Brussels. I came to watch and tend him, as I would have joyfully gone, with such a purpose, to the dreariest ends of the earth. When he knew no one else he knew me. When he lay at the point of death, he married me, that he might call me Wife before he died. And the name, my dear love, that I took on that forgotten night——'

'I know it now! The shadowy remembrance strengthens. It was my name. You are my wife!'

Well! They were happy. It was a long recovery but they were happy through it all. The snow had melted on the ground, and the birds were singing in the early spring, when those three were first able to ride out together, and when people flocked about the open carriage to cheer and congratulate *Captain* Doubledick.

But, even then, it became necessary for the Captain, instead of returning to England, to complete his recovery in the climate of Southern France.

[1] 'Mr. Dickens's voice . . . is not equal to the vocal expression of these fine traits of feminine feeling . . . which his own conceptions evidently demand' (*Wolverhampton Chronicle*, 10 November 1858).

They found a spot upon the Rhone, within a ride of the old town of Avignon, which was all they could desire; and they lived there, together, six months; and then the Captain and his wife returned to England. But Mrs. Taunton growing old and feeling benefited by the change, resolved to remain for a year in those parts. So, she remained: and she was to be rejoined and escorted home, at the year's end, by Captain Doubledick.

She went to the neighbourhood of Aix; and there, *in a chateau near the farmer's house she rented,*[1] she grew into intimacy with a family belonging to that part of France. The intimacy began in her often meeting among the vineyards a pretty child: a girl, who was never tired of listening to the solitary English lady's stories of her poor son and the cruel wars. The family were as gentle as the child, and at length she came to know them so well, that she accepted their invitation to pass the last month of her residence abroad, under their roof. All this intelligence she wrote home, piecemeal as it came about; and, at last, enclosed a polite note from the head of the chateau, soliciting, on the occasion of his approaching mission to that neighbourhood, the honour of the company of cet homme si justement célèbre, Monsieur le Capitaine Richard Doubledick.

Captain Doubledick; now a hardy, handsome man in the full vigour of life, broader across the chest and shoulders than he had ever been before; dispatched a courteous reply, and followed it in person. Travelling through all that extent of country after three years of Peace, he blessed the better days on which the world had fallen. The corn was golden—not drenched in unnatural red; was bound in sheaves for food—not trodden underfoot by men in mortal fight. The smoke rose up from peaceful homes, not blazing ruins. The carts were laden with the fair fruits of the earth, not with wounds and death. To him who had so often seen the terrible reverse, these things were beautiful indeed, and they brought him in *a softened spirit* to the old chateau upon a deep blue evening.

It was a large chateau of the genuine old ghostly kind, with round towers, and extinguishers and a high leaden roof, and more windows than Aladdin's Palace. The lattice blinds were all thrown open, and the entrance doors stood open too—as doors often do in that country when the heat of the day is past; the Captain saw no bell or knocker, and walked in.

He walked into a lofty stone hall, refreshingly cool and gloomy after the glare of a Southern day's travel. Extending along the four sides of this hall, was a gallery, leading to suites of rooms; and it was lighted from the top. Still, no bell was to be seen.

[1] Single continuous underlining.

'Faith,' said the Captain, halting, ashamed of the clanking of his boots, 'this is a ghostly beginning!'

He started back, and felt his face turn white. In the gallery, looking down at him, stood the French officer: the officer whose picture he had carried in his mind so long and so far.

The officer moved, and disappeared, and Captain Doubledick heard his steps coming quickly down into the hall. He entered through an archway. *There was a bright, sudden look upon his face. Such a look as it had worn in that fatal moment.*

Monsieur le Capitaine Richard Doubledick? Enchanted to receive him! A thousand apologies! The servants were all out in the air. There was a little fête among them in the garden. In effect, it was the fête day of my daughter, the little cherished and protected of Madame Taunton.

He was so gracious that Monsieur le Capitaine Richard Doubledick could not withhold his hand. 'It is the hand of a brave Englishman,' said the French officer, retaining it while he spoke. 'I could respect a brave Englishman, even as my foe; how much more as my friend! I, also, am a soldier.'

'He has not remembered me, as I have remembered him; he did not take such note of my face, that day, as I took of his. How shall I tell him!'

The French officer conducted his guest into a garden, and there presented him to his beautiful wife. His daughter came running to embrace him; and there was a boy-baby. A multitude of children-visitors were dancing to sprightly music; and all the servants and peasants about the chateau were dancing too. *It was a scene of innocent happiness that might have been invented for the climax of the scenes of Peace which had soothed the captain's journey.*

He looked on, greatly troubled in his mind, until a resounding bell rang, and the French officer begged to show him his rooms.

'You were at Waterloo?'

'I was. And I was at Badajos.'

Left alone with the sound of his own stern voice in his ears, he sat down to consider, *What shall I do, and how shall I tell him?* At that time, unhappily, many deplorable duels had been fought between English and French officers, arising out of the recent war; and these duels, and how to avoid this officer's hospitality, were the uppermost thought in Captain Doubledick's mind.

He was thinking, and letting the time run out in which he should have dressed for dinner, when Mrs. Taunton spoke to him at his chamber door. '*His mother above all. How shall I tell her?*'[1]

'You will form a friendship with your host, I hope,' said Mrs. Taunton.

[1] *her*, italic in *HW*, is also further underlined in *Berg*.

'He is so true-hearted and so generous, Richard, that you can hardly fail to esteem one another. If He had been spared,' she kissed the locket in which she wore his hair, 'he would have appreciated him with his own generosity, and would have been happy that the evil days were past, which made a man his enemy.'

She left the room; the Captain walked, first to one window whence he could see the dancing in the garden; then, to another window whence he could see the smiling prospect and the peaceful vineyards.

'Spirit of my departed friend,' said he, 'is it through thee, these better thoughts are rising in my mind! Is it thou who hast shown me, all the way I have been drawn to meet this man, the blessings of the altered time! Is it thou who hast sent thy stricken mother to me, to stay my angry hand! Is it from thee the whisper comes, that this man only did his duty as thou didst—and as I did, through thy guidance, which saved me, here on earth—and that he did no more!'

He made the second strong resolution of his life: That neither to the French officer, *nor* to the mother of his departed friend, *nor* to any soul while either of the two was living, would he breathe what only *he* knew. And when he touched that French officer's glass with his own, that day at dinner, he secretly forgave him—forgave him, in the name of the *Divine Forgiver*.

BOOTS AT THE HOLLY-TREE INN

THE second item in Dickens's first programme of short pieces was taken from *The Holly-Tree Inn*, the *Household Words* Christmas number for 1855, and it became one of his most popular: only the *Carol* and *The Trial* received more performances. This story of two little children eloping to Gretna Green may strike the modern reader as coy and embarrassing, and one American critic at the time dismissed it as 'an atrocious yarn', but it greatly touched and amused Dickens's audiences, and George Eliot had found 'the same startling inspiration in his description of "Boots", as in the speeches of Shakespeare's mobs or numbskulls'. Cobbs, the 'Boots', narrates the tale, and his wit and humour offset the sentimentality; moreover, some of the story's more heart-tugging phrases had been omitted from the script. 'The way in which Mr. Dickens read this piece,' reported the *Dublin Evening Mail*, 'is inimitable, and kept his audience convulsed with merriment. . . . The chuckling tone and merry twinkle of the eye with which he gave expression to the salient points of the story were wonderfully effective, and impossible to be described. It was evident that Mr. Dickens entered thoroughly into the spirit of the thing, and enjoyed the fun as heartily as his hearers.' His embodiment of Boots, wrote John Hollingshead, was 'remarkable for ease, finish, and a thorough relish for the character. The swaying to and fro of the body, the half-closing of the eye, and the action of the head, when any point in the narrative is supposed to require particular emphasis to make it clear . . . all assist to make it a perfect example of pure comedy acting' (*Critic*, 4 September 1858).

Cobbs, though working in Yorkshire, had started his working life 'down away by Shooter's Hill there, six or seven miles from Lunnon', and Dickens brought into his conversation

. . . all the characteristics of emphatic humble Cockney conversation, as far as they can be rendered consonant with the graceful interest of the tale. It is quite a treat to hear Mr. Dickens pronounce the Londoner's 'w', which, as those who have nice ears are aware, is only half a 'v', and is grossly caricatured when a full 'v' is substituted for it. The pauses for general emphasis are also the result of very shrewd observation, and the sardonics of Boots are the very cream of universal humour. (*Morning Star*, 6 January 1869)

The 'great hit' of the Reading, treasured by all who heard it, was when Boots 'told of the sympathy of the women with the little pair' (*Scotsman*, 21 December 1868)—their excitement when the children are first installed in their sitting-room, and the female staff of the Inn 'was seven deep at the key-hole', and their anguish later when the children are to be parted and one of the chambermaids calls out, 'It's a shame to part 'em!' It is characteristic of Dickens's adaptations that, in both these places, the rest of the paragraph is deleted, doubtless because it had proved an anti-climax after the 'great hit' (or clap-trap).

The Tale is introduced, and occasionally interrupted, by a traveller who is snowbound in the Holly-Tree Inn. To indicate to himself that these passages were to be read in a different tone and accent, Dickens underlined them and put them inside square brackets, in the prompt-copy (*Berg*). At a few points, he forgets this convention, but in the present reprint consistency has been maintained.

Boots at the Holly-Tree Inn

[1][BEFORE *the days of railways, and in the time of the old great North Road, I was once snowed up at the Holly Tree Inn. Beguiling the days of my imprisonment there, by talking, at one time or other, with the whole establishment, I one day talked with the Boots when he lingered in my room.*]

Where had he been in his time? [*Boots repeated, when I asked him the question.*] Lord, he had been everywhere! [*And what had he been?*] Bless you, everything you could mention a'most.

Seen a good deal? Why, of course he had. [*I should say so, he could assure me, if I only knew about a twentieth part of what had come in his*[2] *way.*] Why, it would be easier for him, he expected, to tell what he hadn't seen, then what he had. Ah! A deal, it would.

What was the curiousest thing he had seen? Well! He didn't know.[3] He couldn't momently name what was the curiousest thing he had seen—unless it was a Unicorn—and he see *him*[4] once at a Fair. But supposing a young gentleman not eight years old, was to run away with a fine young woman of seven, might I think *that*[5] a queer start? Certainly! Then, that was a start as he himself had had his blessed eyes on—and he had cleaned the shoes they run away in—and they was so little that he couldn't get his hand into 'em.

Master Harry Walmers's father, you see, he lived at the Elmses, down away by Shooter's Hill there, six or seven miles from Lunnon. He was a gentleman of spirit, and good looking, and held his head up when he walked, and had what you might call Fire about him. He wrote poetry, and he rode, and he ran, and he cricketed, and he danced, and he acted, and he done it all equally beautiful. He was uncommon proud of Master Harry, as was his only child; but he didn't spoil him, neither. He was a gentleman that had a will of his own and a eye of his own, and that would be minded. Consequently, though he made quite a companion of the fine bright boy, and was delighted to see him so fond of reading his

[1] The first paragraph is added, in manuscript, in the prompt-copy (*Berg*).

[2] *his* italic in *Berg* and in *Household Words* (*HW*).

[3] 'Well! I don't know' (Wright): Dickens doubtless often altered *he* to *I*, when reading, even if not in *Berg*. He was speaking 'in character' here, pronouncing 'curious-est' as 'curious-es-est'. 'Boots . . . approaches the word "curious-es-est" with a look of admiration, clings to every syllable with affection, and only lets go his hold because conversation would otherwise come to a dead lock' (Field, p. 89).

[4] *him* italic in *Berg* and in *HW*. Improved to 'he see him once *in spirits* at a Fair' (Field, p. 89).

[5] *that* italic in *Berg* and in *HW*.

fairy books, and was never tired of hearing him say My name is Norval,[1] or hearing him sing his songs about Young May Moons is a beaming love, and When he as adores thee has left but the name, and that: still he kept the command over the child, and the child *was*[2] a child, and it's wery much to be wished more of[3] 'em was!

[*How did Boots happen to know all this?*] Why, Sir, through being under-gardener. Of course I couldn't be under-gardener, and be always about, in the summer-time, near the windows on the lawn, a mowing and sweeping, and weeding and pruning, and this and that, without getting acquainted with the ways of the family.—Even supposing Master Harry hadn't come to me one morning early, and said, 'Cobbs, how should you spell Norah, if you was asked?' and[4] when I give him my views, Sir, respectin' the spelling o' that name, he took out his little knife, and he begun a cutting it in print, all over the fence.

And the courage of the boy![5] Bless your soul, he'd have throwd off his little hat, and tucked up his little sleeves, and gone in at a Lion, he would. One day he stops, along with her, where I was hoeing weeds in the gravel, and says, speaking up, 'Cobbs,' he says, 'I like *you*.'[6] 'Do you, sir, I'm proud to hear it.' 'Yes I do, Cobbs. Why do I like you, do you think, Cobbs?' 'Don't know, Master Harry, I am sure.' 'Because Norah likes you, Cobbs.' 'Indeed, sir? That's very gratifying.' 'Gratifying, Cobbs? It's better than millions of the brightest diamonds, to be liked by Norah.' 'Certainly, sir.'[7] 'You're going away, ain't you, Cobbs?' 'Yes, sir.' 'Would you like another situation, Cobbs?' 'Well, sir, I shouldn't object, if it was a good 'un.' 'Then, Cobbs,' says that mite, 'you shall be our Head Gardener when we are married.' And he tucks her, in her little sky blue mantle, under his arm, and walks away.

[8][*Boots could assure me that it was*] better than a picter, and equal to a play, to see them babies with their long bright curling hair, their sparkling eyes, and their beautiful light tread, rambling about the garden, deep in love.[9] [*Boots was of opinion*] that the birds believed they was birds, and kept up with 'em, singing to please 'em. Sometimes, they would creep under the Tulip-tree, and would sit there with their arms

[1] This passage was spoken 'monotonously' (Wright).

[2] *was* italic in *Berg* and in *HW*.

[3] '... more *on* 'em ...' (Wright).

[4] 'when I' to 'little knife, and' added in manuscript. Dickens said '... give him my *individual* views ...' (Field, p. 89).

[5] 'Action' (Wright). Dickens's gestures here may be imagined.

[6] *you* italic in *Berg* and in *HW*.

[7] 'Certainly, sir' in a 'Low, quiet voice' (Wright). 'Perhaps, there was after all nothing better in the delivery of the whole of this Reading, than the utterance of [these] two words' (Kent, p. 223).

[8] 'Holding book. Say "Sir" throughout' (Wright).

[9] 'Low monotonous rhythm' (Wright).

round one another's necks, and their soft cheeks touching, a reading about the Prince, and the Dragon, and the good and bad enchanters, and the king's fair daughter. Sometimes I would hear them planning about having a house in a forest, keeping bees and a cow, and living entirely on milk and honey. Once I came upon them by the pond, and heard Master Harry say, 'Adorable Norah, kiss me, and say you love me to distraction, or I'll jump in head foremost.' On the whole, Sir, the contemplation o' them two babbies had a tendency to make me feel as if I was in love myself—only I didn't exactly know who with.

'Cobbs,' says Master Harry, one evening, *when I was watering the flowers*; 'I am going on a visit, this present Midsummer, to my grandmamma's at York.'

'Are you indeed, sir? I hope you'll have a pleasant time. I am going into Yorkshire myself, when I leave here.'

'Are you going to your grandmamma's, Cobbs?'

'No, sir. I haven't got such a thing.'

'Not as a grandmamma, Cobbs?'[1]

'No, sir.'

The boy looks on at the watering of the flowers, for a little while, and then he says, 'I shall be very glad indeed to go, Cobbs—*Norah's going.*'[2]

'You'll be all right then, sir, with your beautiful sweetheart by your side.'

'Cobbs,' returns the boy, a flushing. 'I never let anybody joke about that, when I can prevent them.'

'It wasn't a joke, sir,—wasn't so meant.'

'I am glad of that, Cobbs, because I like you, you know, and you're going to live with us. Cobbs!'

'Sir.'

'What do you think my grandmamma gives me, when I go down there?'

'I couldn't so much as make a guess, sir.'

'A Bank of England five-pound note, Cobbs.'

'Whew![3] that's a spanking sum of money, Master Harry.'

'A person could do a good deal with such a sum of money as that. Couldn't a person, Cobbs?'

'I believe you, sir!'

'Cobbs,' says that boy, 'I'll tell you a secret. At Norah's house, they have been joking her about me, and *pretending to laugh at our being engaged.*[4] Pretending to make game of it, Cobbs!'

[1] 'Harry's voice not high' (Wright).
[2] Doubly underlined.
[3] Instead of saying this word, Dickens whistled (Wright).
[4] 'With a wondering look' (Kent, p. 225).

'Such, sir, is the depravity of human natur.'

The boy, looking exactly like his father, stood for a few minutes, and then departed with 'Good-night, Cobbs. I'm going in.'

[*If I was to ask Boots how it happened*] that I was a going to leave that place at just that present time, well, I couldn't rightly answer you, Sir. I do suppose I might have stayed there till now, if I had been any-ways inclined. But, you see, I was younger then, and I wanted change. That's what I wanted—change. Mr. Walmers, he says to me when I give him notice of my intentions to leave, 'Cobbs,' he says, 'have you anything to complain of? I make the inquiry, because if I find that any of my people really has anythink to complain of, I wish to make it right if I can.' 'No, sir, thanking you, sir, I find myself as well sitiwated here as I could hope to be anywheres. The truth is, sir, that I'm a going to seek my fortun.' 'O, indeed, Cobbs?' he says; 'I hope you may find it.' [*And Boots could assure me—which he did, touching his hair with his bootjack*—that he hadn't found it yet.]

[1]'Well, sir! I left the Elmses when my time was up, and Master Harry he went down to the old lady's at York, which old lady were so wrapt up in that child as she would have give that child the teeth out of her head (*if she had had any*[2]). What does that Infant do—for Infant you may call him and be within the mark—but cut away from that old lady's with his Norah, on a expedition to go to Gretna Green and be married![3]

Sir, I was at this identical Holly-Tree Inn (having left it several times since to better myself, but always come back through one thing or another), when, one summer afternoon, the coach drives up, and out of the coach gets them two children. The Guard says to our Governor, 'I don't quite make out these little passengers, but the young gentle-man's words was, that they was to be brought here.' The young gentleman gets out; hands his lady out; gives the Guard something for himself; says to our Governor, 'We're to stop here to-night, please. Sitting-room and two bed-rooms will be required. Mutton chops and cherry-pudding for two!' and tucks her, in her little sky-blue mantle, under his arm, and walks into the house much bolder than Brass.

Sir, I leave you to judge what the amazement of that establishment was, when those two tiny creatures all alone by themselves was marched into the Angel;—much more so when I, who had seen them without their seeing me, give the Governor my views of the expedition they was upon. 'Cobbs,' says the Governor, 'if this is so, I must set off myself to York and quiet their friends' minds. In which case you must keep

[1] Marginal stage-direction *Piano*.

[2] Doubly underlined. '... her 'ead (*if* she ...' (Wright). 'The h's were perfection' (*Scotsman*, 21 December 1868).

[3] '... to Gretna Green *in Scotland*, and get married' (Wright).

your eye upon 'em, and humour 'em, till I come back. But, before I take these measures, Cobbs, I should wish you to find from themselves whether your opinions is correct.' 'Sir to you,'[1] says I, 'that shall be done directly.'

[*So, Boots goes upstairs to the Angel, and there he finds*] Master Harry on a e-normous sofa—immense at any time, but looking like the Great Bed of Ware, compared with him—a drying the eyes of Miss Norah with his pocket hankecher. Their little legs was entirely off the ground, of course, and it really is not possible to express how small them children looked.

'It's Cobbs! It's Cobbs!' cries Master Harry, and he comes running to me and catching hold of my hand. Miss Norah, she comes running to me on t'other side and catching hold of my t'other hand, and they both jump for joy.

'I see you a getting out, sir,' says I. 'I thought it was you. I thought I couldn't be mistaken in your heighth and figure. What's the object of your journey, sir?—Matrimonial?'

'We are going to be married, Cobbs, at Gretna Green,' returns the boy. 'We have run away on purpose. Norah has been in rather low spirits, Cobbs; but she'll be happy, now we have found you to be our friend.'

'Thank you, sir, and thank *you*,[2] miss, for your good opinion. *Did* you bring any luggage with you, sir?'

[*If I will believe Boots when he gives me his word and honour upon it,*] the lady had got a parasol, a smelling bottle, a round and a half of cold buttered toast, eight peppermint drops, and a Doll's hairbrush. The gentleman had got about half-a-dozen yards of string, a knife, three or four sheets of writing-paper folded up surprisingly small, a orange, and a Chaney mug with his name on it.[3]

'What may be the exact natur of your plans, sir?' says I.

'To go on,' replies the boy—which the courage of that boy was something wonderful!—'in the morning, and be married to-morrow.'

'Just so, sir. Would it meet your views, sir, if I was to accompany you?'

They both jumped for joy again, and cried out, 'O yes, yes, Cobbs! Yes!'

'Well, sir. If you will excuse my having the freedom to give an opinion,

[1] 'Sir to you' was one of 'the little chance phrases, the merest atoms of exclamation here and there, [which] will still be borne in mind as having had an intense flavour of fun about them, as syllabled in the Reading'. Another was Harry's 'Get out with you, Cobbs!' (Kent, p. 227).

[2] *you* and *Did* both italic in *Berg* and in *HW*.

[3] This 'gravely enumerated' account of the children's luggage was one of the most 'irresistibly laughable' passages in the Reading (Kent, p. 226).

what I should recommend would be this. I'm acquainted with a pony, sir, which, put[1] in a pheayton that I could borrow, would take you and Mrs. Harry Walmers Junior (driving myself, if you approved), to the end of your journey in a very short space of time. I am not altogether sure, sir, that this pony will be at liberty till to-morrow, but even if you had to wait over to-morrow for him, it might be worth your while. As to the small account here, sir, in case you was to find yourself running at all short, that don't signify; because I'm a part proprietor of this inn, and it could stand over.'

[*Boots assures me that when they clapped their hands, and jumped for joy again, and called him, 'Good Cobbs!' and 'Dear Cobbs!' and bent across him to kiss one another in the delight of their confiding hearts, he felt himself*] the meanest rascal for deceiving 'em, that was ever born.

'Is there anything you want, just at present, sir?' I says, mortally ashamed of myself.

'We should like some cakes after dinner,' answers Master Harry, 'and two apples—and jam. With dinner we should like to have toast and water. But, Norah has always been accustomed to half a glass of currant wine at dessert. And so have I.'

'It shall be ordered at the bar, sir,' I says.

Sir, I has the feeling as fresh upon me at this minute of speaking, as I had then, that I would far rather have had it out in half-a-dozen rounds with the Governor, than have combined with him; and that I wished with all my heart there was any impossible place where those two babies could make an impossible marriage, and live impossibly happy ever afterwards. However, as it couldn't be, I[2] went into the Governor's plans, and the Governor set off for York in half-an-hour.

The way in which the women of that house—without exception—every one of 'em—married *and*[3] single—took to that boy when they heard the story, is surprising.[4] It was as much as could be done to keep 'em from dashing into the room and kissing him. They climbed up all sorts of places, at the risk of their lives, to look at him through a pane of glass. And they was seven deep at the key-hole.[5]

In the evening, I went into the room to see how the runaway couple was getting on. The gentleman was on the window-seat, supporting the lady in his arms. She had tears upon her face, and was lying, very tired and half asleep, with her head upon his shoulder.

[1] Pronounced as in 'putty' (Wright).

[2] *Berg* here reads 'he went . . .' In the rest of this paragraph, Dickens had altered 'Boots' and 'he' to 'I'; he overlooked this instance. *T & F* corrects it to 'I went . . .'

[3] *and* italic in *Berg* and in *HW*.

[4] Pronounced 'suppurising' (Wright). Boots fondled this word, like 'curious-es-est' (Field, p. 90).

[5] This passage, and the chambermaid's 'It's a shame to part 'em!', 'were always the most telling hits, the chief successes of the Reading' (Kent, p. 229).

'Mrs. Harry Walmers Junior, fatigued, sir?'

[1]'Yes, she is tired, Cobbs; but, she is not used to be away from home, and she has been in low spirits again. Cobbs, do you think you could bring a biffin,[2] please?'

'I ask your pardon, sir. What was it you?—'

'I think a Norfolk biffin would rouse her, Cobbs. She is very fond of them.'

Well, Sir, I withdrew in search of the required restorative, and the gentleman handed it to the lady, and fed her with a spoon, and took a little himself. The lady being heavy with sleep, and rather cross, 'What should you think, sir,' I says, 'of a chamber candlestick?' The gentleman approved; the chambermaid went first, up the great staircase; the lady, in her sky-blue mantle, followed, gallantly escorted by the gentleman; the gentleman embraced her at her door, and retired to his own apartment, where I locked him up.[3]

[*Boots couldn't but feel with increased acuteness what a base deceiver he was, when they consulted him at breakfast* (*they had ordered sweet milk-and-water*, *and toast and currant jelly*, *overnight*), *about the pony. It really was as much as he could do*, *he don't mind confessing to me*,] to look them two young things in the face, and think what a wicked old father of lies he had grown up to be. Howsomever, Sir, I went on a lying like a Trojan, about the pony. I told 'em that it did so unfort'nately happen that the pony was half clipped, you see, and that he couldn't be took out in that state, for fear it should strike to his inside. But that he'd be finished clipping in the course of the day, and that to-morrow morning at eight o'clock the pheayton would be ready. [*Boots's view of the whole case*, *looking back upon it in my room*, *is*,] that Mrs. Harry Walmers Junior was beginning to give in. She hadn't had her hair curled when she went to bed, and she didn't seem quite up to brushing it herself, and its getting in her eyes put her out. But, nothing put out Master Harry. He sat behind his breakfast-cup, a tearing away at the jelly, as if he had been his own father.

In the course of the morning, Master Harry rung the bell—it was surprising how that there boy did carry on—and said in a sprightly way, 'Cobbs, is there any good walks in this neighbourhood?'

'Yes, sir. There's Love Lane.'

'Get out with you, Cobbs!'[4]—that was that there boy's expression—'you're joking.'

'Begging your pardon, sir, there really is Love Lane. And a pleasant

[1] 'Distressed sort of voice' (Wright).

[2] Here, and for 'a Norfolk biffin' below, Dickens substituted 'a baked apple' (Wright; Field, p. 91).

[3] Stressing *locked him up* (Wright).

[4] Spoken quickly (Wright).

walk it is, and proud shall I be to show it to yourself and Mrs. Harry Walmers Junior.'

'Norah, dear,' says Master Harry, 'this is curious. *We*[1] really ought to see Love Lane. Put on your bonnet, my sweetest darling, and we will go there with Cobbs.'

[*Boots leaves me to judge what a Beast he felt himself to be, when that young pair told him, as they all three jogged along together, that they had made up their minds to give him two thousand guineas a year as head gardener, on account of his being so true a friend to 'em.*] Well, sir, I turned the conversation as well as I could, and I took 'em down Love Lane to the water-meadows, and there Master Harry would have drownded himself in a half a moment more, a-getting out a water-lily for her—but nothing daunted that boy. Well, sir, they was tired out. All being so new and strange to 'em, they was tired as tired could be. And they laid down on a bank of daisies, like the children in the wood—leastways meadows[2]—and fell asleep.

[3]I don't know, sir—perhaps you do—why it made a man fit to make a fool of himself, to see them two pretty babies a lying there in the clear still sunny day, not dreaming half so hard when they was asleep, as they done when they was awake. But Lord! when you come to think of yourself, you know, and what a game you have been up to ever since you was in your own cradle, and what a poor sort of chap you are arter all—that's where it is! Don't you see, Sir?

Well, sir, they woke up at last, and then one thing was getting pretty clear to me, namely, that Mrs. Harry Walmerses Junior's temper was on the move. When Master Harry took her round the waist, she said he 'teased her so;' and when he says, 'Norah, my young May Moon, your Harry tease you?' she tells him, 'Yes; and I want to go home!'[4]

A biled fowl, and baked bread-and-butter pudding, brought Mrs. Walmers up a little; but I could have wished, I must privately own to you, sir, to have seen her more sensible of the woice of love, and less abandoning of herself to the currants in the pudding. However, Master Harry he kep up, and his noble heart was as fond as ever. Mrs. Walmers turned very sleepy about dusk, and began to cry. Therefore, Mrs. Walmers went off to bed as per yesterday; and Master Harry ditto repeated.

About eleven or twelve at night, comes back the Governor in a chaise, along with Mr. Walmers and a elderly lady. Mr. Walmers says to our missis, 'We are much indebted to you, ma'am, for your kind care of our little children, which we can never sufficiently acknowledge. Pray

[1] *We* doubly underlined.

[2] The interjected 'leastways meadows' was spoken quickly (Wright).

[3] The paragraph was spoken 'Feelingly and monotonous[ly]' (Wright).

[4] 'Quick and snappishly' (Wright).

ma'am, where is my boy?' Our missis says, 'Cobbs has the dear child in charge, sir. Cobbs, show Forty!' Then, Mr. Walmers, he says: 'Ah, Cobbs! I am glad to see *you.*[1] I understood you was here!' And I says, 'Yes, sir. Your most obedient, sir. I beg your pardon, sir,' I adds, while unlocking the door; 'I hope you are not angry with Master Harry. For, Master Harry is a fine boy, sir, and will do you credit and honour.' [*And Boots signifies to me, that if the fine boy's father had contradicted him in the state of mind in which he then was, he thinks he should have*] 'fetched him a crack,' and took the consequences.

But, Mr. Walmers only says, 'No, Cobbs. No, my good fellow. Thank you!' And, the door being opened, goes in, goes up to the bedside, bends gently down, and kisses the little sleeping face. Then, he stands looking at it for a minute, looking wonderfully like it (*they do say he ran away with Mrs. Walmers*); and then he gently shakes the little shoulder.

'Harry, my dear boy! Harry!'

Master Harry starts and looks at his Pa. Looks at me too. Such is the honour of that mite, that he looks at me, to see whether he has brought me into trouble.

'I am not angry, my child. I only want you to dress yourself and come home.'

'Yes, Pa.'

Master Harry dresses himself quick.

'Please may I'—the spirit of that little creatur!—'please, dear Pa—may I—kiss Norah, before I go?'

'You may, my child.'

So he takes Master Harry in his hand, and I leads the way with the candle, to that other bedroom: where the elderly lady is seated by the bed, and poor little Mrs. Harry Walmers Junior is fast asleep. There, the father lifts the boy up to the pillow, and he lays his little face down for an instant by the little warm face of poor little Mrs. Harry Walmers Junior, and gently draws it to him—a sight so touching to the chambermaids who are a peeping through the door, that one of them calls out 'It's a shame to part 'em!'[2]

[*Finally, Boots says,*] that's all about it. Mr. Walmers drove away in the chaise, having hold of Master Harry's hand. The elderly lady and Mrs. Harry Walmers Junior that was never to be (she married a Captain, long afterwards, and died in India), went off next day. [*In conclusion, Boots puts it to me whether I hold with him in two opinions:*] firstly, that there are not many couples on their way to be married, who are half as innocent as them two children; secondly, that it would be a jolly

[1] *you* italic in *Berg* and in *HW*.

[2] 'High female voice' (Wright); 'shrill' (Kent, p. 230). 'May the exclamation of the soft-hearted chambermaid ... never vanish from my memory!' (A. W. Ward, *Dickens* (1882), p. 152). See headnote, above.

good thing for a great many couples on their way to be married, if they could only be stopped in time and brought back separate.[1]

[1] 'With which cynical scattering of sugar-plums in the teeth of married and single, the blithe Reading was laughingly brought to its conclusion' (Kent, p. 230). A friend of Dickens, Frederick Lehmann, after attending a performance of *Boots* in 1859, wrote to his wife that it was 'a perfect masterpiece as a reading, and certainly that charming little tale "*se prête admirablement*" to such a performance. It was curious though', he added, referring to the well-publicized breakup of Dickens's marriage the year before, 'to hear him wind up with [this] very significant moral' (quoted by John Lehmann, *Ancestors and Friends* (1962), p. 164). Dickens had elsewhere identified himself with the quizzical views of 'my friend the Boots' (*N*, ii, 736–7), quoting as his own sentiments Boots's 'when you come to think of yourself you know . . .' (see previous page but one).

MRS. GAMP

Mrs. Gamp, which concluded the programme of three items first performed on 17 June 1858, was much the longest, and it gave Dickens the most trouble. This was only the second Reading he had devised from a novel, and his principle of selection was to base it upon one character—Mrs. Gamp being, of course, one of the most admired of all his creations. The initial process of adaptation was considerably simpler than on comparable occasions later, when he was more experienced. A text was prepared in two chapters. Chapter I was a slightly abbreviated version of Mrs. Gamp's first appearance in *Martin Chuzzlewit* (ch. 19): she is fetched by Mr. Pecksniff to lay out the body of Anthony Chuzzlewit; there are dialogues involving Mr. Mould, Jonas Chuzzlewit, and old Chuffey, and the chapter ends with Mrs. Gamp's feeling disposed to put her lips to several glasses of liquor and thus dropping off into somnolent silence. A few Gampisms from chs. 46 and 40 were interpolated into this narrative. Chapter II was taken entirely from Mrs. Gamp's next entry in the novel (ch. 25): she calls upon Mr. Mould, and after pleasant conversation with him goes to the Bull in Holborn, where she relieves Betsey Prig, who is nursing Mr. Lewsome through a fever. Mrs. Gamp settles down for the night, but is disturbed by her patient's strange and incriminating ejaculations during his delirium. The Reading ended with Mrs. Prig's returning in the morning; Mrs. Gamp departs, commending the cucumbers.

The reception of *Mrs. Gamp* was very mixed, critics disagreeing over this item more than any other. It was 'the strong point of the evening' or its lowspot; Dickens's performance was brilliant—or it had fallen entirely flat. Certainly he seems not to have hit a consistent form in the crucial role of Mrs. Gamp, and some of his other characterizations were found disappointing by well-disposed critics. He was indeed tempting Providence by undertaking such legendary characters as Mrs. Gamp, or Mr. Pecksniff who makes a brief and unimpressive entry in this item: as over Sam Weller in *The Trial*, audiences had such high expectations and strong preconceptions that almost any performance was likely to disconcert or disappoint. Nevertheless there was much praise for his facial expressions as Mrs. Gamp and his voice for her, 'snuffy, husky, unctuous, the voice of a fat old woman' (Kent, p. 212).

One criticism—that 'there is not much point in the story'—Dickens recognized and acted upon. The original version was inconsequential and ended feebly. Eventually, Dickens halved its length, jettisoned Chapter II (though salvaging some of its best detachable passages), kept all the action in the Chuzzlewit house (substituting Chuffey for Lewsome as the patient), and wrote a new and better conclusion, raiding ch. 49 for the famous quarrel between Betsey and Mrs. Prig over Mrs. Harris. This process of revision can be followed in the facsimile of the 1858 prompt-copy (*Berg*) edited by John D. Gordan (New York Public Library, 1956); as Gordan remarks, none of Dickens's Readings was so drastically rewritten.

After the 1858–9 tours, *Mrs. Gamp* was performed only infrequently, and in America it was the least often performed item in the repertoire. During the 1868–9 Farewell series, however, Dickens's fondness for it revived and he often used it as an afterpiece or in another much-performed triple bill—*Boots*, *Sikes and Nancy* and *Mrs. Gamp*. Around this time he made a new prompt-copy (*Starling*)—reprinted below, as the latest version—in a copy of the 1867–8 American published booklet (*T & F*). Its differences from *Berg* are slight. A third prompt-copy, described by Kent (p. 141), seems not to have survived; this is the only prompt-copy known to have been in existence after Dickens's death which has not been available to the present editor.

Mrs. Gamp

Mr. Pecksniff was in a hackney-cabriolet, for Jonas Chuzzlewit had said, 'Spare no expense.' It should never be charged upon his father's son that he grudged the money for his father's funeral.

Mr. Pecksniff had been to the undertaker, and was now on his way to another officer in the train of mourning,—a female functionary, a nurse, and watcher, and performer of nameless offices about the persons of the dead,—whom the undertaker had recommended. Her name was Gamp; her residence, in Kingsgate Street, High Holborn. So Mr. Pecksniff, in a hackney-cab, was rattling over Holborn's stones, in quest of Mrs. Gamp.

This lady[1] lodged at a bird-fancier's, *next door but one to the celebrated mutton-pie shop, and directly opposite to the original cat's-meat warehouse.* It was a little house, and this was the more convenient; for Mrs. Gamp being also a monthly nurse, and lodging in the first-floor front, her window was easily assailable at night by pebbles, walking-sticks, and fragments of tobacco-pipe,—all much more efficacious than the street-door knocker; *which was so ingeniously constructed as to wake the street with ease, without making the smallest impression on the premises to which it was addressed.*

It chanced on this particular occasion that Mrs. Gamp had been up *with a distressed Lady* all the previous night. It chanced that Mrs. Gamp had not been regularly engaged, but had been called in at a crisis, to assist another professional lady with her advice; and thus it happened that, all points of interest in the case being over, Mrs. Gamp had come home again to the bird-fancier's, and gone to bed. So, when Mr. Pecksniff drove up in the hackney-cab, Mrs. Gamp's curtains were drawn close, and Mrs. Gamp was fast asleep behind them. †

Mr. Pecksniff, in the innocence of his heart, applied himself to the knocker; *but at his first double-knock, every window in the street became alive with female heads*;[2] and before he could repeat it, whole troops of married ladies came flocking round the steps, all crying out with one accord, and with uncommon interest, 'Knock at the winder, sir, knock at the winder. Lord bless you, don't lose no more time than you can help,—knock at the winder!'

Borrowing the driver's whip for the purpose, Mr. Pecksniff soon made

[1] 'This *excellent* lady' (Wright).

[2] Dickens stressed '*alive*', and his eyes became 'so distended at the extraordinary spectacle as to remove all doubt as to the possibility of such a commotion' (Field, p. 128).

a commotion among the first-floor flower-pots, and roused Mrs. Gamp, whose voice—to the great satisfaction of the matrons—was heard to say, 'I'm coming.'

'He's as pale as a muffin,' said one lady, in allusion to Mr. Pecksniff.

'So he ought to be, if he's the feelings of a man,' observed another.[1]

A third lady said she wished he had chosen any other time for fetching Mrs. Gamp, but it always happened so with *her*.[2]

It gave Mr. Pecksniff much uneasiness to infer, from these remarks, that he was supposed to have come to Mrs. Gamp *upon an errand touching, not the close of life, but the other end.* Mrs. Gamp herself was under the same impression, for, throwing open the window, she cried behind the curtains, as she hastily dressed herself:

'Is it Mrs. Perkins?'

'No! nothing of the sort.'

'What, Mr. Whilks! Don't say it's you, Mr. Whilks, and that poor creetur Mrs. Whilks with not even a pincushion ready. Don't say it's you, Mr. Whilks!'[3]

'It isn't Mr. Whilks. I don't know the man. Nothing of the kind. A gentleman is dead; and some person being wanted in the house, you have been recommended by Mr. Mould the undertaker. You are also wanted to relieve Mrs. Prig, the day-nurse in attendance on the book-keeper of the deceased,—one Mr. Chuffey,—whose grief seems to have affected his mind.'[4]

As she was by this time in a condition to appear, Mrs. Gamp, who had a face for all occasions, looked out of the window with her mourning countenance, and said she would be down directly.

But the matrons took it very ill that Mr. Pecksniff's mission was of so unimportant a kind; and rated him in good round terms, signifying that they would be glad to know what he meant by terrifying delicate females 'with his corpses', and giving it as their opinion that he was ugly enough to know better.

So, when Mrs. Gamp appeared, the unoffending gentleman was glad

[1] 'Nodding head' (Wright).

[2] *her* italic in the novel. The three ladies were 'defined with photographic accuracy': the first, 'of measured medium voice and scrutinizing eye'; the second 'of nervous-sanguine temperament', speaking quickly and with a toss of the head; the third 'of a melancholy turn of mind and cast of countenance, the born victim of circumstances' (Field, p. 129).

[3] Kate Field describes this as an 'exclamation', but then corrects herself: 'There is an intellectual ponderosity about her that renders an exclamation impossible. She carries too much ballast. . . . She scorns staccato passages, and her vocalisation may be said to be confined to the use of semi-breves, on which she lingers as if desirous of developing her voice by what is technically known as "swelling". She holds all notions of light and shade in contempt, and with monotonous cadence produces effects upon her hearers undreamed of by her readers' (p. 130).

[4] This sentence (not in the novel) is written into *Berg*.

to hustle her into the cabriolet, and drive off, *overwhelmed with popular execration.*

Mrs. Gamp had a large bundle with her, a pair of pattens, and a species of gig umbrella; the latter article in color like a faded leaf, except where a circular patch of a lively blue had been let in at the top. She was much flurried by the haste she had made, and labored under the most erroneous views of cabriolets, which she appeared to confound with mail-coaches or stage-wagons, insomuch that she was constantly endeavoring for the first half-mile to force her luggage through the little front window, and clamoring to the driver to 'put it in the boot.' When she was disabused of this idea, her whole being resolved itself into an absorbing anxiety about her pattens, *with which she played innumerable games of quoits on Mr. Pecksniff's legs.* It was not until they were close upon the house of mourning that she had enough composure to observe:—

'And so the gentleman's dead, sir! Ah! The more's the pity,'—*she didn't even know his name.* 'But it's what we must all come to. It's as certain as being born, except that we can't make our calculations as exact. Ah! Poor dear!'

She was a fat old woman, with a husky voice and a moist eye. Having very little neck, it cost her some trouble to look over herself, if one may say so, at those to whom she talked. She wore a rusty black gown, rather the worse for snuff, and a shawl and bonnet to correspond. * *The face of Mrs. Gamp—the nose in particular—was somewhat red and swollen, and it was difficult to enjoy her society without becoming conscious of a smell of spirits.*

'Ah!' repeated Mrs. Gamp, *for that was always a safe sentiment in cases of mourning,*—'ah, dear! When Gamp was summonsed to his long home, and I see him a lying in the hospital with a penny-piece on each eye, and his wooden leg under his left arm, I thought I should have fainted away. But I bore up.'

If certain whispers current in the Kingsgate Street circles had any truth in them, Mrs. Gamp had borne up surprisingly, *and had indeed exerted such uncommon fortitude as to dispose of Mr. Gamp's remains for the benefit of science.*

'You have become indifferent since then, I suppose? Use is second nature, Mrs. Gamp.'

'You may well say second natur, sir. One's first ways is to find sich things a trial to the feelings, and such is one's lasting custom.[1] If it wasn't for the nerve a little sip of liquor gives me (which I was never able to do more than taste it), I never could go through with what I sometimes has to do. "Mrs. Harris," I says, at the wery last case as ever

[1] The spellings *sich, such* and *sech* all occur in the Reading text.

I acted in, which it was but a young person,[1]—"Mrs. Harris," I says, "leave the bottle on the chimley-piece, and don't ask me to take none, but let me put my lips to it when I am so dispoged, and then I will do what I am engaged to do, according to the best of my ability." "Mrs. Gamp," she says, in answer, "if ever there was a sober creetur to be got at eighteen-pence a day for working people, and three and six for gentlefolks,—*night watching being a extra charge*—you are that inwallable person." "Mrs. Harris," I says to her, "don't name the charge, for if I could afford to lay all my fellow-creeturs out for nothink, I would gladly do it, sich is the love I bears 'em."'[2]

At this point, she was fain to stop for breath. And advantage may be taken of the circumstance, *to state that a fearful mystery surrounded this lady of the name of Harris*, whom no one in the circle of Mrs. Gamp's acquaintance had ever seen; neither did any human being know her place of residence. The prevalent opinion was that she was a phantom of Mrs. Gamp's brain, created for the purpose of holding complimentary dialogues with her on all manner of subjects.

'The bottle shall be duly placed on the chimney-piece, Mrs. Gamp, and you shall put your lips to it at your own convenience.'

'Thank you, sir. Which it is a thing as hardly ever occurs with me, unless when I am indispoged, and find my half a pint o' porter settling heavy on the chest. Mrs. Harris often and often says to me, "Sairey Gamp," she says, "you raly do amaze me!" "Mrs. Harris," I says to her, "why so? Give it a name, I beg!" "Telling the truth then, ma'am," says Mrs Harris, "and shaming him as shall be nameless betwixt you and me, never did I think, till I know'd you, as any woman could[3] sick-nurse and monthly likeways, on the little that you takes to drink." "Mrs. Harris," I says to her, "none on us knows what we can do till we tries; and wunst *I*[4] thought so, too. But now," I says, "my half a pint of porter fully satisfies; perwisin', Mrs. Harris, that it is[5] brought reg'lar, and draw'd mild."'

The conclusion of this affecting narrative brought them to the house.

[1] In *Berg*, the words 'which it was but a young person' are underlined. Wright notes here: 'Head moving from side to side'.

[2] 'The expression of her glowing face at this juncture defies language, however live, particularly as she remarks to Mrs. Harris, with a pendulum wag to her head in the *tempo* of a funeral march, "If I could possible afford to lay . . . [*etc.*]' (Field, p. 134). Maybe Dickens inserted a 'possible' here: or maybe Kate Field was misquoting from memory.—The next three paragraphs are an interpolation, made in *Berg* when the Reading was shortened; the first and third paragraphs were transferred from Chapter II of the *Berg* text (ch. 25 of the novel), and the middle paragraph was a newly written bridge-passage.

[3] *could* doubly underlined in *Berg*.

[4] *I* italic in *Starling*.

[5] *is* doubly underlined in *Berg*. Wright notes an additional phrase here: Dickens read '. . . perwisin', Mrs. Harris, *which I makes a condition*, that it is . . .'

In the passage they encountered Mr. Mould, the undertaker, a little elderly gentleman, bald, and in a suit of black; with a note-book in his hand, and a face in which a queer attempt at melancholy was at odds with a smirk of satisfaction.

'Well, Mrs. Gamp, and how are *you*,[1] Mrs. Gamp?'

'Pretty well, I thank you, sir.'

'You'll be very particular here, with the deceased party upstairs, Mrs. Gamp. This is not a common case, Mrs. Gamp. Let everything be very nice and comfortable, about the deceased, Mrs. Gamp, if you please.'[2]

'It shall be so, sir; you knows me of old, I hope, and so does Mrs. Mould, your ansome pardner, sir; and so does the two sweet young ladies, your darters; although the blessing of a daughter was deniged myself, which, if we had had one, Gamp would certainly have drunk its little shoes right off its feet, as with our precious boy he did, and arterwards send the child a errand, to sell his wooden leg *for any liquor it would fetch as matches in the rough*; which was truly done beyond his years, for ev'ry individgle penny that child lost at tossing for kidney-pies, and come home arterwards quite[3] bold, to break the news, *and offering to drown himself if sech would be a satisfaction to his parents*. But wery different is them two sweet young ladies o' yourn, Mister Mould, as I know'd afore a tooth in their pretty heads was cut, and have many a time seen—ah! the dear creeturs!—a playing at berryin's down in the shop, and a follerin' the order-book to its long home in the iron safe. Young ladies with such faces as your darters thinks of something else besides berryin's; don't they, sir? Thinks o' marryin's; don't they, sir?'

'I'm sure I don't know, Mrs. Gamp. * Very shrewd woman, Mr. Pecksniff, sir. Woman whose intellect is immensely superior to her station in life; sort of woman one would really almost feel disposed to bury for nothing, and do it neatly, too. Mr. Pecksniff, sir. This Funeral is one of the most impressive cases, sir, that I have seen in the whole course of my professional experience.'

'Indeed, Mr. Mould!'

'Such affectionate regret I never saw. There is no limitation; there is positively NO[4] limitation in point of expense! I have orders, sir, in short, to turn out something absolutely gorgeous.'

'My friend Mr. Jonas is an excellent man.'

[1] *you* italic in *Starling* and in the novel.

[2] The next two paragraphs (up to Mr. Mould's 'and do it neatly, too, Mr. Pecksniff, sir') were interpolated into *Berg*, being a rearranged version of some passages from the discarded Chapter II (ch. 25 of the novel).

[3] Here Ticknor, to whom Dickens gave this prompt-copy, inserted, with a caret, 'sweet and', and noted: 'So read by C.D.'

[4] Small capitals in *Starling* and in the novel.

'Well, I have seen a good deal of what is filial in my time, sir, and of what is unfilial, too! It is the lot of parties in my line, sir. We come into the knowledge of those secrets. But anything so filial as this—anything so honorable to human nature, anything so expensive, so calculated to reconcile all of us to the world we live in—never yet came under my observation. It only proves, sir, what was so forcibly observed by the lamented poet,[1]—*buried at Stratford*,—that there is good in everything.'

'It is very pleasant to hear you say so, Mr. Mould.'

'You are very kind, sir. And what a man the late Mr. Chuzzlewit was, sir! Ah! what a man he was. * Mr. Pecksniff, sir, good morning!'

Mr. Pecksniff returned the compliment; and Mould was going away with a brisk smile, when he remembered the occasion. Quickly becoming depressed again, he sighed; *looked into the crown of his hat, as if for comfort; put it on*[2] *without finding any; and slowly departed.*

Mrs. Gamp and Mr. Pecksniff then ascended the staircase; and Mrs. Gamp, having been shown to the chamber in which all that remained of old Anthony Chuzzlewit lay covered up, with but one loving heart, and that the heart of his old book-keeper, to mourn it, left Mr. Pecksniff free to enter the darkened room below in search of Mr. Jonas.

He found that example to bereaved sons, and pattern in the eyes of all performers of funerals, so subdued, that he could scarcely be heard to speak, and only seen to walk across the room.[3]

'Pecksniff, you shall have the regulation of it all, mind! You shall be able to tell anybody who talks about it, that everything was correctly and freely done. There isn't any one you'd like to ask to the funeral, is there?'

'No, Mr. Jonas, I think not.'

'Because if there is, you know, ask him. We don't want to make a secret of it.'

'No; I am not the less obliged to you on that account, Mr. Jonas, for your liberal hospitality; but there really is no one.'

'Very well; then you, and I, and the doctor, will be just an easy coachful. We'll have the doctor, Pecksniff, because he knows what was the matter with my father, and that it couldn't be helped.' *

With that, they went up to the room where the old book-keeper was, attended by Mrs. Betsey Prig. And to them entered Mrs. Gamp soon

[1] In *Berg*, it is 'lamented theatrical poet'; Dickens failed to notice that 'theatrical' had been omitted in *T & F*, but (as Wright notes) he said the omitted word.

[2] 'Put on hat' (Wright).

[3] 'The brief glimpse of Jonas Chuzzlewit gave an opportunity for the display of the highest dramatic power. The suspicious glance; the morose, churlish voice; the incessant biting of the thumb-nail, betrayed the conscious parricide' (*Belfast News-Letter*, 30 August 1858).

afterwards, who saluted Mrs. Prig as one of the sisterhood, and 'the best of creeturs.'[1]

The old book-keeper sat beside the bed, with his head bowed down; until Mrs. Gamp took him by the arm, when he meekly rose, saying:—[2]

'My old master died at threescore and ten,—ought and carry seven. Some men are so strong that they live to fourscore—four times ought's an ought, four times two's an eight—eighty. Oh! why—why—why—didn't he live to four times ought's an ought, and four times two's an eight—eighty? Why did he died before his poor old crazy servant! Take him from me, and what remains? I loved him. He was good to me. I took him down once, six boys, in the arithmetic class at school. God forgive me! Had I the heart to take him down!'

[3]'Well I'm sure,' said Mrs. Gamp, 'you're a wearing old soul, and that's the blessed truth. You ought to know that you was born in a wale, and that you live in a wale, and that you *must take the consequences of sich a sitivation.* As a good friend of mine has frequent made remark to me, Mr. Jonage Chuzzlewit, which her name, sir,—I will not deceive you,—is Harris,—Mrs. Harris through the square and up the steps a turnin' round by the tobacker shop,—and which she said it the last Monday evening as ever dawned upon this Pilgrim's Progress of a mortal wale,[4]—"O Sairey, Sairey, little do we know wot lays afore us!" "Mrs. Harris, ma'am," I says, "not much, it's true, but more than you suppoge. Our calcilations, ma'am," I says, "respectin' wot the number of a family will be, comes most times within one, and oftener than you would suppoge, exact." "Sairey," says Mrs. Harris, in a awful way, "tell me wot is my individgle number." "No, Mrs. Harris," I says to her, "ex-cuge me, if you please. My own family," I says, "has fallen out of three-pair backs, and has had damp doorsteps settled on their lungs, and one was turned up smilin' in a bedstead unbeknown.[5] Therefore, ma'am," I says, "seek not to protigipate, but take 'em as they come and as they

[1] This paragraph was written in, in *Berg*, where the next two pages were wafered together. It draws on ch. 25.

[2] 'Chuffey . . . stands out vividly. . . . The fine "points" of this short monologue are seized by Dickens. The picture of the meek, heart-broken, maundering, faithful servant, with decrepit figure, quavering voice, and trembling hands, . . . is painted in natural colors; nor is there exaggeration in the drawing' (Field, p. 131).

[3] The following speech is made up of Gampisms from several chapters; it was printed here in *Berg*, but Dickens interpolated, in manuscript, the phrases about being 'born in a wale' and the 'Pilgrim's Progress' (*sic: Berg* has the compromise 'Pilgian's Progress', but *Starling* regularizes 'Piljian's Projiss' entirely). Wright alters it to 'Projiss'. The interpolations come from chs. 46, 49 and 40 (pp. 705, 751, 625).

[4] 'There is a sibylline tendency in her look as she ecstatically gazes towards heaven' at this point (Field, p. 136).

[5] 'Here Mrs. Gamp suits the action to the word and smiles the smile of confiding youth and innocence. Its appeal is irresistible' (Field, p. 135).

go. Mine," I says to her,—"mine is all gone, my dear young chick.[1] And as to husbands, there's a wooden leg gone likewise home to its account, which, in its constancy of walking into public-'ouses, and never coming out again till fetched by force, was quite as weak as flesh, if not weaker."'[2]

Mrs. Gamp, now left to the live part of her task, formally relieved Mrs. Prig for the night. That interesting lady had a gruff voice and a beard, and straightway got her bonnet and shawl on. *

'Anythink to tell afore you goes, Betsey, my dear?'

'The pickled salmon in this house is delicious. I can partickler recommend it. The drinks is all good. His physic and them things is on the drawers and mankleshelf. He took his last slime draught at seven. The easy-chair ain't soft enough. You'll want his piller.'[3]

Mrs. Gamp thanked Mrs. Prig for these friendly hints and gave her good night. † She then entered on her official duties: firstly, she put on a yellow-white nightcap of prodigious size, in shape resembling a cabbage: *having previously divested herself of a row of bald old curls, which could scarcely be called false, they were so innocent of anything approaching to deception*;[4] secondly, and lastly, she summoned the housemaid, to whom she delivered this official charge, in tones expressive of faintness:—

'I think, young woman, as I could peck a little bit o' pickled salmon, with a little sprig o' fennel, and a sprinkling o' white pepper. I takes new bread, my dear, with jest a little pat o' fredge butter and a mossel o' cheese. With respect to ale, if they draws the Brighton old tipper at any 'ouse nigh here, I takes that[5] ale at night, my love; not as I cares for it myself, but on accounts of it being considered wakeful by the doctors; and whatever you do, young woman, don't bring me more than a shillingsworth of gin-and-water, warm, when I rings the bell a second time; for that is always my allowange, and I never takes a drop beyond. In case there should[6] be sich a thing as a cowcumber in the 'ouse, I'm rather partial to 'em, though I am but a poor woman. *Rich folks may ride on camels, but it ain't so easy for them to see out of a needle's eye.* That is my comfort, and I hopes I knows it.'

[1] In *Berg*, the last four words are underlined.

[2] This is the penultimate paragraph of Chapter I of the *Berg* text. At this point, Dickens interpolated two closely-written manuscript pages, and then directed himself to a further page mostly made up from stuck-in extracts from ch. 49 (about Mrs. Gamp's quarrel with Mrs. Prig.). These three pages constituted the revised ending of the Reading, replacing the thirty-three pages of Chapter II: but most of the two manuscript pages were versions of passages salvaged from that Chapter.

[3] Voice too mannish: a demoralized Dickens, not really a Betsey Prig (Field, p. 133).

[4] According to Charles Kent (p. 216), Dickens later omitted the address to the housemaid, which follows here.

[5] In *Berg*, Dickens underlined 'that': also the next phrase from 'not as I cares for it' to 'by the doctors'.

[6] In *Berg*, 'should' is underlined.

The supper and drink being brought and done full justice to, she administered the patient's medicine by the simple process of clutching his windpipe to make him gasp, and immediately pouring it down his throat.

'Drat the old wexagious creetur,[1] I a'most forgot your piller, I declare!' she said, drawing it away. 'There! Now you're as comfortable as you need be, I'm sure! and *I*'m[2] a going to be comfortable too.'

All her arrangements made, she lighted the rushlight, coiled herself up on her couch, and fell asleep.

Ghostly and dark the room, and full of shadows. The noises in the streets were hushed, the house was quiet, the dead of night was coffined in the silent city. * When Mrs. Gamp awoke, she found that the busy day was broad awake too. Mrs. Prig relieved punctually, having passed a good night at another patient's. But Mrs. Prig relieved in an ill temper.

The best among us have their failings,[3] and it must be conceded of Mrs. Prig, that if there were a blemish in the goodness of her disposition, it was a habit she had of not bestowing all its sharp and acid properties upon her patients (*as a thoroughly amiable woman would have done*), but of keeping a considerable remainder for the service of her friends. * She looked offensively at Mrs. Gamp, and winked her eye. Mrs. Gamp felt it necessary that Mrs. Prig should know her place, and be made sensible of her exact station in society. * So she began a remonstrance with:—

'Mrs. Harris, Betsey—'

'Bother Mrs. Harris!'[4]

Mrs. Gamp looked at Betsey with amazement, incredulity, and indignation. Mrs. Prig, winking her eye tighter, folded her arms and uttered these tremendous words:—

'I don't believe there's no sich a person!'[5]

With these expressions, she snapped her fingers once, twice, thrice, each time nearer to the face of Mrs. Gamp, and then turned away as one who felt that there was now a gulf between them which nothing could ever bridge across.

[1] 'Drat the old wexagious creetur,' inserted in manuscript in *Starling*, is a conflation of two phrases from ch. 46 (pp. 705, 709).

[2] *I* italic in *Starling*. In *Berg*, 'you' in 'you need' is underlined.

[3] This passage was spoken by 'the Reader with a hardly endurable gravity of explanation' (Kent, p. 218). This final episode comes from ch. 49.

[4] 'Sneeringly' (Wright).

[5] 'Shaking head' (Wright).

BARDELL AND PICKWICK

The Trial from 'Pickwick', as this item was generally called, was first performed on 19 October 1858: 'I took it into my head' to read it, Dickens said (*N*, iii. 64). It had not been announced, but obviously it had been thoroughly prepared, and it became a very popular afterpiece; the *Carol* and *The Trial* was his most-used single programme and, as was noted in the *Carol* headnote above, was his constant stand-by for important occasions. Indeed, *The Trial* was his most frequently performed item. All but a few phrases come from *Pickwick Papers*, ch. 34, here reduced by about a half—too severe a truncation, some thought—and Dickens assumed that his audiences needed no reminder of how Mr. Pickwick had given Mrs. Bardell grounds for her action.

Since 1837, this had been a favourite comic episode, often adapted for the stage, and 'done to death' in Penny-readings and other such performances. Every public reader, amateur or professional, had it in his repertoire. Dickens's rendering topped everyone's, it was generally reported, both in narrative powers and in characterizations less hackneyed and more credible than those of platform-tradition:

Those public readers or actors . . . who have read or performed this scene have unavoidably given it an air of burlesque or farce . . . Mr. Dickens, with the privilege of the author, has done what no one else has ventured to do. He has omitted, added, and altered, and has in this way succeeded in giving the scene an air of probability which in the original version it does not wear. The greatest change has been in the character of Mr. Winkle, who in the original is represented as a weak, timid and almost idiotic young man, but who in Mr. Dickens's new version gets into something like a passion with the opposing barrister, and shows a certain amount of courage and resolution. The humour is still exaggerated but it no longer runs riot with excess of caricature. (*Bath Chronicle*, 14 February 1867)

Mr. Justice Stareleigh was a great hit: Dickens adopted a marvellously comic face, used an equally comic voice (variously described as sepulchral and drawling), and seemed to take on the physical shape of the fat little judge:

Dickens steps out of his own skin which, for the time being, is occupied by Mr. Justice Stareleigh. His little round eyes, wide open and blinking; his elevated eyebrows that are in a constant state of interrogation; his mouth, drawn down by the weight of the law; the expression of the *ensemble*, which clearly denotes that everybody *is* a rascal whether found guilty or not; and the stern, iron-clad voice, apparently measuring out justice in as small quantities as possible, and never going faster than a dead march,—make up an impersonation that is extraordinary, even for Dickens. (Field, p. 103)

Most of the eight characters were much relished and admired. As Professor Adolphus Ward exclaimed:

Talk to us of Garrick and his Protean powers! Mr. Justice Stareleigh, and Mr. Serjeant Buzfuz, and Mr. Nathaniel Winkle himself, all appeared in succession in Mr. Dickens's face and voice, and, like the image of Mrs. Bardell's departed exciseman, will remain 'stamped on' us for ever. If Mr. Samuel Weller failed to attain to a similar incarnation, it is only because, like his own personal illustrations, he has already become to all of us a mythical personage incapable of realisation in this imperfect world. (*Manchester Guardian*, 4 February 1867)

The one disappointment, indeed, was Sam Weller, for the reason most picturesquely expressed by the Yankee who stalked out of a performance because, he said, Dickens knew 'no more about Sam Weller 'n a cow does of pleatin' a shirt, at all events that ain't *my* idea of Sam Weller, anyhow' (Dolby, p. 176). Everyone had his own delighted notion of Sam, who was indeed, as Ward suggests, too 'mythical' a character not to be a disappointment in the flesh. The one indubitable success associated with Sam was the applause, even 'roar of delight', which greeted the first mention of his name. As a critic remarked, on Dickens's New York début, 'It was such an unaffected tribute of admiration as few authors have ever obtained. Mr. Dickens stood before us in the flesh—listening to that voice of human sympathy and admiration which only the posterity of most great men hear.' (*Nation*, 12 December 1867)

The 1858 prompt-copy has not survived, and Dickens's one private copy which has done so (*Suzannet*, where it is bound with two 1861 pieces, *Mr. Chops, the Dwarf* and *Mr. Bob Sawyer's Party*), contains no underlinings or performance-signs. Happily, W. M. Wright records Dickens's gestures, etc., very thoroughly, in a volume discovered too late to be used in my annotation. Buzfuz ('too slow and tame', Wright thought) turned indirect into direct speech—e.g., 'the responsibeeleety imposed *upon* me'—and was idiosyncratic in his pronunciations. He pronounced the plaintiff's name 'Bardill' and, when reading the letters, said 'Mistress B'. Her placard read '*In*quire with*in*'. On 'The plaintiff is a widow', he 'Brushes tear from right eye with handkerchief and crushes tear from left eye with left hand.' He 'slapped book' on 'Chops!', 'sauce' and at other points, and further amplified his peroration to '. . . a contemplative, and I am persuaded I'm not going too far in saying, an eminently practical, and a highly philosophical and poetical, jury . . .' Dickens omitted 'inquired a juror' and instead leant forward. Stareleigh wrote in his book on 'Daniel' and elsewhere, sometimes spoke 'sleepily', and said Winkle would be 'sent to moulder in gaol' instead of 'committed to prison'. Winkle stuttered and stammered, leaning forward and speaking very rapidly; at 'I am not intimate' he spoke 'Crossly, rising on toes and coming down by way of emphasis' and at Pickwick's 'holding the plaintiff in his arms' he gulped and paused. Mr. Skimpin, dramatizing 'the occasion in question', put 'Arm out as if enclosing waist'. Every time Dickens spoke as Sam Weller he 'leant forward on right arm on table'.

Bardell and Pickwick

On the morning of the trial of the great action for breach of promise of marriage—Bardell against Pickwick—the defendant, Mr. Pickwick, being escorted into court, stood up in a state of agitation, and took a glance around him. There were already a pretty large sprinkling of spectators in the gallery, and a numerous muster of gentlemen in wigs, in the barristers' seats: who presented, as a body, all that pleasing and extensive variety of nose and whisker for which the bar of England is justly celebrated. Such of the gentlemen as had a brief to carry, carried it in as conspicuous a manner as possible, and occasionally scratched their noses with it, to impress it more strongly on the observation of the spectators. Other gentlemen who had no briefs, carried under their arms goodly octavos, with a red label behind, and that under-done-pie-crust-coloured cover, which is technically known as 'law calf.' Others, who had neither briefs nor books, thrust their hands into their pockets, and looked as wise as they could. The whole, to the great wonderment of Mr. Pickwick, were divided into little groups, who were chatting and discussing the news of the day in the most unfeeling manner possible,—just as if no trial at all were coming on. *

A loud cry of 'Silence!' announced the entrance of the judge: who was most particularly short, and so fat, that he seemed all face and waistcoat. He rolled in, upon two little turned legs, and having bobbed to the bar, who bobbed to him, put his little legs underneath his table, and his little three-cornered hat upon it.[1] * A sensation was then perceptible in the body of the court; and immediately afterwards Mrs. Bardell, the plaintiff, supported by Mrs. Cluppins, her bosom friend number one, was led in in a drooping state. An extra-sized umbrella was then handed in by Mr. Dodson, and a pair of pattens by Mr. Fogg (Dodson and Fogg being the plaintiff's attornies), each of whom had prepared a sympathizing and melancholy face for the occasion. Mrs. Sanders, bosom friend number two, then appeared, leading in Master Bardell, * whom she placed on the floor of the court in front of his hysterical mother,—a commanding position in which he could not fail to awaken the sympathy of both judge and jury. This was not done without considerable opposition on the part of the young gentleman himself, who had misgivings that his being placed in the full glare of the judge's eye was only a formal prelude to his being immediately ordered away for instant execution.

[1] *Suzannet* reads '... hat upon it; a sensation ...' *1883* correctly begins a new sentence here. *Suzannet* is punctuated thus because, after a long deletion (the swearing-in of the jury) the text resumes in the middle of a sentence.

'I am for the plaintiff, my Lord,' said Mr. Serjeant Buzfuz.

COURT.—'Who is with you, brother Buzfuz?' Mr. Skimpin bowed, to intimate that he was.

'I appear for the defendant, my Lord,' said Mr. Serjeant Snubbin.

COURT.—'Anybody with you, brother Snubbin?'

'Mr. Phunky, my Lord.' *

COURT.—'Go on.'

Mr. Skimpin proceeded to 'open the case;' and the case appeared to have very little inside it when he had opened it, for he kept such particulars as he knew, completely to himself.

Serjeant Buzfuz then rose with all the majesty and dignity which the grave nature of the proceedings demanded, and having whispered to Dodson, and conferred briefly with Fogg, pulled his gown over his shoulders, settled his wig, and addressed the jury.

Serjeant Buzfuz began by saying,[1] that never, in the whole course of his professional experience—never, from the very first moment of his applying himself to the study and practice of the law—had he approached a case with such a heavy sense of the responsibility imposed upon him—a responsibility he could never have supported, were he not buoyed up and sustained by a conviction so strong, that it amounted to positive certainty that the cause of truth and justice, or, in other words, the cause of his much-injured and most oppressed client, *must* prevail[2] with the high-minded and intelligent dozen of men whom he now saw in that box before him.

Counsel always begin in this way, because it puts the jury on the best terms with themselves, and makes them think what sharp fellows they must be. A visible effect was produced immediately; several jurymen beginning to take voluminous notes.

'You have heard from my learned friend, gentlemen,' continued Serjeant Buzfuz, well knowing that, from the learned friend alluded to, the gentlemen of the jury had heard nothing at all—'you have heard from my learned friend, gentlemen, that this is an action for a breach of promise of marriage, in which the damages are laid at 1500*l.* But

[1] 'The oration of the learned Serjeant . . . was a triumph in its way—Mr. Dickens ingeniously contriving to accent every word that the rules of elocution would forbid to be accented' (*Scotsman*, 21 April 1866). Charles Kent writes of Buzfuz's 'extraordinarily precise, almost mincing pronunciation'; for instance, he pronounced *responsibility* 'respon-see-bee-lee-ty' (pp. 112–13). Kate Field remarks on his 'rising inflection', and uses a quasi-musical notation to indicate the pitch and stress of various phrases (pp. 103–7). Wright annotates Buzfuz extensively: see p. 122.

[2] The word *must*, like all words italicized in this Reading, is printed italic in *Suzannet*. Dickens would repeat the phrase 'must prevail', and 'The intonation of these final words are delightfully burlesque. Serjeant Buzfuz draws back his head and then throws it forward to add impressiveness to speech, while a muscular contortion going on at the back of his neck and rippling down his shoulders suggests memories of a heavy swell on the ocean' (Field, pp. 104–5).

you have not heard from my learned friend, inasmuch as it did not come within my learned friend's province to tell you, what are the facts and circumstances of this case. Those facts and circumstances, gentlemen, you shall hear detailed by me, and proved by the unimpeachable female whom I will place in that box before you.[1] The plaintiff is a widow; yes, gentlemen, a widow. The late Mr. Bardell, after enjoying, for many years, the esteem and confidence of his sovereign, as one of the guardians of his royal revenues, glided almost imperceptibly from the world, to seek elsewhere for that repose and peace which a custom-house can never afford.'

This was a pathetic description of the decease of Mr. Bardell, who had been knocked on the head with a quart-pot in a public-house cellar.

'Some time before Mr. Bardell's death, he had stamped his likeness on a little boy.[2] With this little boy, the only pledge of her departed exciseman, Mrs. Bardell shrunk from the world, and courted the retirement and tranquillity of Goswell Street; and here she placed in her front parlour-window a written placard, bearing this inscription—"Apartments furnished for a single gentleman. Inquire within."' Here Serjeant Buzfuz paused, while several gentlemen of the jury took a note of the document.

'There is no date to that, is there, sir?' inquired a juror.

'There is no date, gentlemen, but I am instructed to say that it was put in the plaintiff's parlour-window just this time three years. Now, I entreat the attention of the jury to the wording of this document—"Apartments furnished for a single gentleman!" "Mr. Bardell," said the widow; "Mr. Bardell was a man of honour—Mr. Bardell was a man of his word—Mr. Bardell was no deceiver—Mr. Bardell was once a single gentleman himself; *in* single gentlemen I shall perpetually see something to remind me of what Mr. Bardell was, when he first won my young and untried affections; to a single gentleman shall my lodgings be let." Actuated by this beautiful and touching impulse, (among the best impulses of our imperfect nature, gentlemen,) the desolate widow dried her tears, furnished her first floor, caught her innocent boy to her maternal bosom, and put the bill up in her parlour-window. Did it remain there long? No. Before the bill had been in the parlour-window three days—three days, gentlemen—a Being, erect upon two legs, and bearing all the outward semblance of a man, and not of a monster, knocked at Mrs. Bardell's door. He inquired within; he took the lodgings; and on the very next day he entered into possession of them. This man was Pickwick—Pickwick the defendant.'

Serjeant Buzfuz here paused for breath. The silence awoke Mr. Justice

[1] Here 'Buzfuz' gave a mighty thump on the desk, as the novel text indicates (Field, p. 105).

[2] Pronounced '*lit*-tle bo—*hoy*' (Field, p. 105).

Stareleigh, who immediately wrote down something with a pen without any ink in it, and looked unusually profound, to impress the jury with the belief that he always thought most deeply with his eyes shut.

'Of this man Pickwick I will say little; the subject presents but few attractions; and I, gentlemen, am not the man, nor are you, gentlemen, the men, to delight in the contemplation of revolting heartlessness, and of systematic villany.'

Here Mr. Pickwick, who had been writhing in silence, gave a violent start, as if some vague idea of assaulting Serjeant Buzfuz, in the august presence of justice and law, suggested itself to his mind.

'I say systematic villany, gentlemen,' said Serjeant Buzfuz, looking through Mr. Pickwick, and talking *at* him; 'and when I say systematic villany, let me tell the defendant Pickwick—if he be in court, as I am informed he is[1]—that it would have been more decent in him, more becoming, in better judgment, and in better taste, if he had stopped away.[2] * I shall show you, gentlemen, that for two years Pickwick continued to reside without interruption or intermission, at Mrs. Bardell's house. I shall show you that, on many occasions, he gave halfpence, and on some occasions even sixpences, to her little boy; and I shall prove to you, by a witness whose testimony it will be impossible for my learned friend to weaken or controvert, that on one occasion he patted the boy on the head, and, after inquiring whether he had won any *alley tors* or *commoneys* lately (both of which I understand to be a particular species of marbles much prized by the youth of this town), made use of this remarkable expression—"How should you like to have another father?" I shall prove to you, gentlemen, * on the testimony of three of his own friends—most unwilling witnesses, gentlemen—most unwilling witnesses—that on that morning he was discovered by them holding the plaintiff in his arms, and soothing her agitation by his caresses and endearments.[3] And now, gentlemen, but one word more. Two letters have passed between these parties—letters which are admitted to be in the handwriting of the defendant. * Let me read the first:—"Garraway's, twelve o'clock. Dear Mrs. B.—Chops and Tomata sauce. Yours, PICKWICK." Gentlemen, what does this mean? Chops! Gracious heavens! and Tomata[4] sauce! Gentlemen, is the happiness of a sensitive and confiding female to be trifled away by such shallow artifices as these? The next has no date whatever, which is in itself suspicious.—"Dear Mrs. B., I shall not be at home till to-morrow. Slow coach." And then follows

[1] Rising to a *crescendo* on 'he is'. 'When Serjeant Buzfuz . . . [here] aims a forefinger at the defendant's head, it becomes a query, whether grotesque action is not as difficult to excel in as absolute grace. Dickens has learned its secret' (Field, p. 106).

[2] *T & F* begins a new paragraph here, as does the novel.

[3] *T & F* again begins a new paragraph here, as does the novel.

[4] Spelled *Tomato* in *T & F*, throughout.

this very remarkable expression—"Don't trouble yourself about the warming-pan." Why, gentlemen, who *does* trouble himself about a warming-pan? Why is Mrs. Bardell so earnestly entreated not to agitate herself about this warming-pan, unless it is, as I assert it to be, a mere cover for hidden fire—a mere substitute for some endearing word or promise, agreeably to a preconcerted system of correspondence, artfully contrived by Pickwick with a view to his contemplated desertion, and which I am not in a condition to explain?[1] * Enough of this: my client's hopes and prospects are ruined. * But Pickwick, gentlemen, Pickwick, the ruthless destroyer of this domestic oasis in the desert of Goswell Street—Pickwick, who has choked up the well, and thrown ashes on the sward—Pickwick, who comes before you to-day with his heartless Tomata sauce and warming-pans—Pickwick still rears his head with unblushing effrontery, and gazes without a sigh on the ruin he has made. Damages, gentlemen—heavy damages—are the only punishment with which you can visit him; the only recompense you can award to my client. And for those damages she now appeals to an enlightened, a high-minded, a right-feeling, a conscientious, a dispassionate, a sympathising, a contemplative[2] jury of her civilized countrymen.'

With this beautiful peroration, Mr. Serjeant Buzfuz sat down, and Mr. Justice Stareleigh woke up.

'Call Elizabeth Cluppins,' said Sergeant Buzfuz, rising a minute afterwards, with renewed vigour. *

'Do you recollect, Mrs. Cluppins? Do you recollect being in Mrs. Bardell's back one pair of stairs, on one particular morning in July last, when she was dusting Pickwick's apartment?'

'Yes, my Lord and Jury, I do.'

'Mr. Pickwick's sitting-room was the first floor front, I believe?'

'Yes, it were, sir.'

COURT.—'What were you doing in the back room, ma'am?'[3]

'My Lord and Jury, I will not deceive you.'

COURT.—'You had better not, ma'am.'

'I was there, unbeknown to Mrs. Bardell; I had been out with a little basket, gentlemen, to buy three pound of red kidney purtaties, which was three pound tuppence ha'penny, when I see Mrs. Bardell's street door on the jar.'

COURT.—'On the what?'[4]

'Partly open, my Lord.'

[1] Again, *T & F* begins a new paragraph here, as does the novel.

[2] Amplified by Dickens to 'and, it is not going too far to say, a highly poetical jury . . .' (New York *Nation*, 12 December 1867, p. 482). See also Wright, p. 122, above.

[3] Mr. Justice Stareleigh's inquiry was suspicious; Mrs. Cluppins replied lackadaisically, and Stareleigh's 'You had better not, ma'am' was very fierce (Kent, p. 114).

[4] Spoken 'in a state of owl-like astonishment' (Field, p. 108).

COURT.—'She *said* on the jar.'

'It's all the same, my Lord.'

The little judge looked doubtful, and said he'd make a note of it.

'I walked in, gentlemen, just to say good mornin', and went, in a permiscuous manner, up stairs, and into the back room. Gentlemen, there was the sound of voices in the front room, and—'

'And you listened, I believe, Mrs. Cluppins?'

'Beggin' your pardon, sir, I would scorn the haction. The voices was very loud, sir, and forced themselves upon my ear.'

'Well, Mrs. Cluppins, you were not listening, but you heard the voices. Was one of those voices, Pickwick's?'

'Yes, it were, sir.'

And Mrs. Cluppins, after distinctly stating that Mr. Pickwick addressed himself to Mrs. Bardell, repeated, by slow degrees, and by dint of many questions, the conversation she had heard. Which, like many other conversations repeated under such circumstances, or indeed like many other conversations repeated under any circumstances, was of the smallest possible importance in itself, but looked big now.[1]

Mrs. Cluppins having broken the ice, thought it a favourable opportunity for entering into a short dissertation on her own domestic affairs; so, she straightway proceeded to inform the court that she was the mother of eight children at that present speaking, and that she entertained confident expectations of presenting Mr. Cluppins with a ninth, somewhere about that day six months. At this interesting point, the little judge interposed most irascibly, and the worthy lady was taken out of court.

'Nathaniel Winkle!' said Mr. Skimpin.

'Here!'[1] Mr. Winkle entered the witness-box, and having been duly sworn, bowed to the judge: who acknowledged the compliment by saying:

COURT.—'Don't look at me, sir; look at the jury.'

Mr. Winkle obeyed the mandate, and looked at the place where he thought the jury might be.

Mr. Winkle was then examined by Mr. Skimpin.[2]

'Now, sir, have the goodness to let his Lordship and the jury know what your name is, will you?' Mr. Skimpin inclined his head on one

[1] Sentence added by Dickens, maybe recalling rumours about his marital troubles.

[2] '"He-ah, he-ah", replies an embarrassed voice' (Field, p. 108). 'Don't we remember how, even before he could open his lips, he was completely disconcerted?' (Kent, p. 116). Mr. Winkle's speech was 'lively and nervously rapid', in contrast to Dickens's narrative pace, which was 'generally slow and measured' (*Western Daily Mercury*, 7 January 1862). There was also a happy contrast between 'the sepulchral tone' of Mr. Justice Stareleigh and what the *Scotsman* (21 April 1866) described as Winkle's 'sharp curt speech'. See also Wright, above, p. 122.

[3] Mr. Skimpin remained 'as vividly as anything at all about this Reading in our recollection' (Kent, p. 117). During his badgering of Mr. Winkle, 'the expression of his countenance denoted positive delight in the work before him' (Field, p. 108).

side and listened with great sharpness for the answer, as if to imply that he rather thought Mr. Winkle's natural taste for perjury would induce him to give some name which did not belong to him.

'Winkle.'

COURT.—'Have you any Christian name, sir?'[1]

'Nathaniel, sir.'

COURT.—'Daniel,—any other name?'

'Nathaniel, Sir—my Lord, I mean.'

COURT.—'Nathaniel Daniel, or Daniel Nathaniel?'

'No, my Lord, only Nathaniel—not Daniel at all.'

COURT.—'What did you tell me it was Daniel for then, sir?'

'I didn't, my Lord.'

COURT.—'You did, sir. How could I have got Daniel on my notes, unless you told me so, sir?'

'Mr. Winkle has rather a short memory, my Lord; we shall find means to refresh it before we have quite done with him, I dare say. Now, Mr. Winkle, attend to me, if you please, sir; and let me recommend you be careful. I believe you are a particular friend of Pickwick, the defendant, are you not?'

'I have known Mr. Pickwick now, as well as I recollect at this moment, nearly—'

'Pray, Mr. Winkle, do not evade the question. Are you, or are you not, a particular friend of the defendant's?'

'I was just about to say, that—'

'Will you, or will you not, answer my question, sir?'[2]

COURT.—'If you don't answer the question, you'll be committed to prison, sir.'

'Yes, I am.'

'Yes, you are. And couldn't you say that at once, sir? Perhaps you know the plaintiff too—eh, Mr. Winkle?'

'I don't know her; but I've seen her.'

'Oh, you don't know her, but you've seen her? Now, have the goodness to tell the gentlemen of the jury what you mean by *that*, Mr. Winkle.'

'I mean that I am not intimate with her, but that I have seen her when I went to call on Mr. Pickwick, in Goswell Street'[3]

'How often have you seen her, sir?'

[1] This improvement on the novel—where the Judge had asked 'What's your Christian name, sir?'—is typical of many small changes made before this text was printed. Mr Justice Stareleigh's 'slow, authoritative tone, as if founded on the Rock of Ages' (Field, p. 109), during this exchange, was much admired.

[2] Here Winkle was given an extra speech: 'Why, God bless my soul, I was just about to say that—' (Kent, p. 111).

[3] Winkle preluded this speech too with 'God bless my soul!' (Field, p. 110).

'How often?'

'Yes, Mr. Winkle, how often? I'll repeat the question for you a dozen times, if you require it, sir.'

On this question there arose the edifying brow-beating, customary on such points. First of all, Mr. Winkle said it was quite impossible for him to say how many times he had seen Mrs. Bardell. Then he was asked if he had seen her twenty times, to which he replied, 'Certainly,—more than that.' Then he was asked whether he hadn't seen her a hundred times—whether he couldn't swear that he had seen her more than fifty times—whether he didn't know that he had seen her at least seventy-five times—and so forth.

'Pray, Mr. Winkle, do you remember calling on the defendant Pickwick at these apartments in the plaintiff's house in Goswell Street, on one particular morning, in the month of July last?'

'Yes, I do.'

'Were you accompanied on that occasion by a friend of the name of Tupman, and another of the name of Snodgrass?'

'Yes, I was.'

'Are they here?'

'Yes, they are,' looking very earnestly towards the spot where his friends were stationed.

'Pray attend to me, Mr. Winkle, and never mind your friends,' with an expressive look at the jury. 'They must tell their stories without any previous consultation with you, if none has yet taken place (another look at the jury). Now, sir, tell the gentlemen of the jury what you saw on entering the defendant's room, on this particular morning. Come; out with it, sir; we must have it, sooner or later.'

'The defendant, Mr. Pickwick, was holding the plaintiff in his arms, with his hands clasping her waist, and the plaintiff appeared to have fainted away.'

'Did you hear the defendant say anything?'

'I heard him call Mrs. Bardell a good creature, and I heard him ask her to compose herself, for what a situation it was, if anybody should come, or words to that effect.'

'Now, Mr. Winkle, I have only one more question to ask you. Will you undertake to swear that Pickwick, the defendant, did not say on the occasion in question, "My dear Mrs. Bardell, you're a good creature; compose yourself to this situation, for to this situation you must come," or words to *that* effect?'

'I—I didn't understand him so, certainly. I was on the staircase, and couldn't hear distinctly; the impression on my mind is—'

'The gentlemen of the jury want none of the impressions on your mind, Mr. Winkle, which I fear would be of little service to honest straightforward men. You were on the staircase, and didn't distinctly hear; but you will not swear that Pickwick did not make use of the expressions I have quoted? Do I understand that?'

'No I will not.' *

'You may leave the box, sir.'

Tracy Tupman, and Augustus Snodgrass, were severally called into the box; both corroborated the testimony of their unhappy friend; and each was driven to the verge of desperation by excessive badgering.

Susannah Sanders was then called, and examined by Serjeant Buzfuz, and cross-examined by Serjeant Snubbin. Had always said and believed that Pickwick would marry Mrs. Bardell; knew that Mrs. Bardell's being engaged to Pickwick was the current topic of conversation in the neighbourhood, after the fainting in July. Had heard Pickwick ask the little boy how he should like to have another father. Did not know that Mrs. Bardell was at that time keeping company with the baker, but did know that the baker was then a single man and is now married. Thought Mrs. Bardell fainted away on the morning in July, because Pickwick asked her to name the day; knew that she (witness) fainted away stone dead when Mr. Sanders asked *her* to name the day, and believed that anybody as called herself a lady would do the same under similar circumstances. During the period of her keeping company with Mr. Sanders she had received love letters, like other ladies. In the course of their correspondence Mr. Sanders had often called her a 'duck,' but he had never called her 'chops,' nor yet 'tomata sauce.'[1]

Serjeant Buzfuz now rose with more importance than he had yet exhibited, if that were possible, and said 'Call Samuel Weller.'[2]

It was quite unnecessary to call Samuel Weller; for Samuel Weller stepped into the box the instant his name was pronounced; and placing his hat on the floor, and his arms on the rail, took a bird's-eye view of the bar, and a comprehensive survey of the bench with a remarkably cheerful and lively aspect.

COURT.—'What's your name, sir?'

'Sam Weller, my Lord.'

COURT.—'Do you spell it with a "V" or with a "W"?'[3]

'That depends upon the taste and fancy of the speller, my Lord. I never had occasion to spell it more than once or twice in my life, but I spells it with a "V."'

[1] Strangely, Dickens omits the next sentences, which provide the pay-off to this joke: 'He was particularly fond of ducks. Perhaps if he had been as fond of chops and tomata sauce, he might have called her that, as a term of affection.' Mr. Emlyn Williams, in his version, retains them, in abbreviated form. Mr. Williams's version, about the same length as Dickens's, is superior to it in retaining more joke-lines and a better continuity; his abbreviations, however, damage the prose-rhythms more than Dickens's do.

[2] See above, p. 122, for the applause which invariably greeted this line.

[3] On the niceties of Cockney *v/w* sounds, see above p. 97.

Here a voice in the gallery exclaimed, 'Quite right too, Samivel; quite right. Put it down a we, my Lord, put it down a we.'

COURT.—'Who is that, who dares to address the court? Usher.'

'Yes, my Lord.'

COURT.—'Bring that person here instantly.'

'Yes, my Lord.'

But as the usher didn't find the person, he didn't bring him; and, after a great commotion, all the people who had got up to look for the culprit, sat down again. The little judge turned to the witness as soon as his indignation would allow him to speak, and said,

COURT.—'Do you know who that was, sir?'

'I rayther suspect it was my father, my Lord.'

COURT.—'Do you see him here now?'

Sam stared up into the lantern in the roof of the court, and said, 'Why no, my Lord, I can't say that I *do* see him at the present moment.'

COURT.—'If you could have pointed him out, I would have sent him to jail instantly.'

Sam bowed his acknowledgments.

'Now, Mr. Weller,' said Serjeant Buzfuz.

'Now, sir.'

'I believe you are in the service of Mr. Pickwick, the defendant in this case. Speak up, if you please, Mr. Weller.'

'I mean to speak up, sir. I am in the service o' that 'ere gen'l'man, and a wery good service it is.'

'Little to do, and plenty to get, I suppose?'

'Oh, quite enough to get, sir, as the soldier said ven they ordered him three hundred and fifty lashes.'

COURT.—'You must not tell us what the soldier said, unless the soldier is in court, and is examined in the usual way; it's not evidence.'[1]

'Wery good, my Lord.'

'Do you recollect anything particular happening on the morning when you were first engaged by the defendant; eh, Mr. Weller?'

'Yes I do, sir.'

'Have the goodness to tell the jury what it was.'

'I had a reg'lar new fit out o' clothes that mornin', gen'l'men of the jury, and that was a wery partickler and uncommon circumstance vith me in those days.'

[1] The novel does not contain 'unless the soldier is in court, and is examined in the usual way'. Dickens further amplified this famous moment: sometimes, apparently, making the Judge inquire whether the soldier was in court (*Berwick and Kelso Warder*, 29 November 1861), sometimes restoring from the novel the phrase '... what the soldier, *or any other man*, said ...' (*Northern Whig*, 21 March 1867), sometimes specifying that 'You cannot possibly be allowed to inform the Court' (etc.) unless the soldier 'is in court, in full Regimentals' (Wright).

The judge looked sternly at Sam, but Sam's features were so perfectly serene that the judge said nothing.

'Do you mean to tell me, Mr. Weller, that you saw nothing of this fainting on the part of the plaintiff in the arms of the defendant, which you have heard described by the witnesses?'

'Certainly not, sir, I was in the passage 'till they called me up; and then the old lady as you call the plaintiff, she warn't there, sir.'

'You were in the passage and yet saw nothing of what was going forward. Have you a pair of eyes, Mr. Weller?'

'Yes, I have a pair of eyes, and that's just it. If they wos a pair o' patent double million magnifyin' gas microscopes of hextra power, p'r'aps I might be able to see through two flights o' stairs and a deal door; but bein' only eyes, you see, my wision's limited.' *

'Now, Mr. Weller, I'll ask you a question on another point, if you please.'

'If you please, sir.'

'Do you remember going up to Mrs. Bardell's house, one night in November?'

'Oh yes, wery well.'

'Oh, you *do* remember that, Mr. Weller. I thought we should get at something at last.'

'I rayther thought that, too, sir.'

'Well; I suppose you went up to have a little talk about the trial—eh, Mr. Weller?'

'I went up to pay the rent; but we *did* get a talkin' about the trial.'

'Oh, you did get a talking about the trial. Now what passed about the trial; will you have the goodness to tell us, Mr. Weller?'

'Vith all the pleasure in life, sir. Arter a few unimportant observations from the two wirtuous females as has been examined here to-day, the ladies gets into a wery great state o' admiration at the honourable conduct of Mr. Dodson and Mr. Fogg—them two gen'l'men as is settin' near you now.'

'The attornies for the plaintiff. Well, they spoke in high praise of the honourable conduct of Messrs. Dodson and Fogg, the attornies for the plaintiff, did they?'

'Yes; they said what a wery gen'rous thing it was o' them to have taken up the case on spec, and not to charge nothin' at all for costs, unless they got 'em out of Mr. Pickwick.'

'It's perfectly useless, my Lord, attempting to get at any evidence through the impenetrable stupidity of this witness. I will not trouble the court by asking him any more questions. Stand down, sir. * That's my case, my Lord.'

Serjeant Snubbin then addressed the jury on behalf of the defendant; *

and did the best he could for Mr. Pickwick; and the best, as everybody knows, could do no more.

Mr. Justice Stareleigh summed up, in the old-established form. He read as much of his notes to the jury as he could decipher on so short a notice; he didn't read as much of them as he couldn't make out; and he made running comments on the evidence as he went along. If Mrs. Bardell were right, it was perfectly clear Mr. Pickwick was wrong; and if they thought the evidence of Mrs. Cluppins worthy of credence, they would believe it; and, if they didn't, why they wouldn't.[1] The jury then retired to their private room to talk the matter over, and the judge retired to *his* private room to refresh himself with a mutton chop and a glass of sherry.

An anxious quarter of an hour elapsed; the jury came back; and the judge was fetched in. Mr. Pickwick put on his spectacles, and gazed at the foreman.

'Gentlemen, are you all agreed upon your verdict?'

'We are.'

'Do you find for the plaintiff, gentlemen, or for the defendant?'

'For the plaintiff.'

'With what damages, gentlemen?'

'Seven hundred and fifty pounds.'

Mr. Pickwick having drawn on his gloves with great nicety, and stared at the foreman all the while, * allowed himself to be assisted into a hackney-coach, which had been fetched for the purpose by the ever-watchful Sam Weller.

Sam had put up the steps, and was preparing to jump on the box, when he felt himself gently touched on the shoulder; and his father stood before him.

'Samivel! The gov'nor ought to have been got off with a alleybi. Ve got Tom Vildspark off o' that 'ere manslaughter (that come of hard driving) vith a alleybi, ven all the big vigs to a man, said as nothing couldn't save him.[2] I know'd what 'ud come o' this here way o' doin' bisniss. O Sammy, Sammy, vy worn't there a alleybi!'

THE END OF BARDELL AND PICKWICK

[1] *T & F* begins a new paragraph here.

[2] These sentences are adapted from ch. 33, but the Reading adds the bracketed admission about Tom Wildspark's culpability.

DAVID COPPERFIELD

In the summer of 1861, having finished *Great Expectations*, Dickens began to think about his new series of Readings, due to start in October. He had given very few performances since his first tour ended in February 1859, and the last new item to enter his repertoire had been *The Trial*, in October 1858. He had been very busy meanwhile with establishing his new weekly *All the Year Round* (April 1859) and writing *A Tale of Two Cities* and *Great Expectations* for serialization in it. He now needed, therefore, to re-rehearse his old repertoire and he decided also to amplify it. This summer, he devised no less than six new items—*Copperfield* and *Nicholas Nickleby*, which opened his autumn tour, *Mr. Bob Sawyer's Party*, which entered the repertoire in December, *Mr. Chops, the Dwarf*, from a Christmas story, which remained unperformed for some years, and two items which were never performed, *The Bastille Prisoner* from *A Tale of Two Cities*, and a very ambitious narrative covering most of *Great Expectations*.

Back in 1855 he had considered varying his Charity Readings of the *Carol* with a *Copperfield* piece to be called 'Young Housekeeping and Little Emily', telling 'the story of David's married life with Dora, and the story of Mr. Peggotty's search for his niece', but had found these episodes very difficult to extract from their intricately woven context (*N*, ii. 619). He returned to the task in 1858 and 1859, again without success, but in 1861 devised a script very much as originally envisaged. It was his most ambitious exercise in selection and revision yet undertaken; the earlier Readings derived from his novels had concentrated on a single character (*Mrs. Gamp*) or episode (*The Trial*) or a limited stretch of the story (*Little Dombey*). He soon realized that he had made a false start; he devised a five-chapter script (a facsimile was published in 1922, ed. J. H. Stonehouse) but then decided that this opened too abruptly, so he added a preliminary chapter (in which Steerforth is introduced to the Peggotty family, and obviously begins to form designs upon Emily) and had a new prompt-copy printed (*Berg*). He then worked intensively on the text, well over a third of which was eventually cut: only nine of its 120 pages are left unamended. During the autumn of 1861, it was a two-hour item; then, in January 1862, he further shortened and revised the text, which was henceforth performed with an afterpiece, commonly *Bob Sawyer*. After all these cuts, the Reading contains some awkward transitions, over-rapid through compression, such as Mr. Peggotty's narrative and plans at the beginning of Chapter VI, after an eight-page cut. Also the 'Courtship and Young Housekeeping' chapters, though delightful, make little sense in relation to the Reading's main plot, the Steerforth–Emily–Peggotty story. Here Dickens obtains contrast and variety at the expense of unity, in what he had called the 'streaky-bacon' narrative technique (*Oliver Twist*, ch. 17) and had indicated in his plans for *David Copperfield*, No. X: 'First chapter funny Then on *to Emily*.'

Like the novel from which it was taken, this Reading was Dickens's favourite. His most exhausting, too: it took so much out of him, he said, that he was half dead the day after. It culminated in the big emotional and physical effort of the great storm at Yarmouth—a passage admired by many critics as the finest thing he wrote and by audiences as the most sublime moment in all the Readings. This item contained a dozen characters, in which he was said to have excelled, and ranged from the comedy (with touches of pathos) centring on David and Dora, and of Micawber, to the drama and tragedy of the main plot, where Mr. Peggotty in his anguish reminded one critic of King Lear, while Steerforth's dark destructiveness was 'as magically suggestive as anything imagined of Faust'. Dickens's letters show him rejoicing in audiences and individuals specially appreciative of his force and of his finer points in this item—the great actor Macready, for instance, who was almost speechless with admiration:

'No—er—Dickens! I swear to Heaven that, as a piece of passion and playfulness—er—indescribably mixed up together, it does—er—no, really, Dickens!—amaze me as profoundly as it moves me. But as a piece of art—and you know—er—that I—no, Dickens! By——! have seen the best art in a great time—it is incomprehensible to me. How is it got at—er—how is it done—er—how one man can—well? It lays me on my—er—back, and it is of no use talking about it!' With which he put his hand upon my breast and pulled out his pocket-handkerchief, and I felt as if I were doing somebody to his Werner. (*N*, iii. 276–7)

Nor was Macready just being polite: the Reading was admirable, he wrote in his diary, 'The humour was delightful, and the pathos of various passages gave me a choking sensation . . . Altogether a truly artistic performance' (Sir Nevil Macready, *Annals of an Active Life* (1924), i. 20). Nearly half a century after Dickens's death, Lord Redesdale wrote: 'Never shall I forget the effect produced by his reading of the death of Steerforth; it was tragedy itself, and when he closed the book and his voice ceased, the audience for a moment seemed paralysed, and one could almost hear a sigh of relief' (*Memories* (1915), ii. 518). Thackeray's daughter Anny (Lady Ritchie), who saw the final performance of *Copperfield* (1 March 1870), also recalled this storm scene as more thrilling than anything she had ever seen in a theatre, and wondered at the power of this 'slight figure (so he appeared to me)' to command a huge audience 'in some mysterious way . . . It was not acting, it was not music, nor harmony of sound and colour, and yet I have the impression of all these things as I think of that occasion' (*From the Porch* (1913), pp. 43–4).

David Copperfield

CHAPTER I[1]

I had known Mr. Peggotty's house very well in my childhood, and I am sure I could not have been more charmed with it, if it had been Aladdin's palace, roc's egg and all. It was an old black barge or boat, high and dry on Yarmouth Sands, with an iron funnel sticking out of it for a chimney. There was a delightful door cut in the side, and it was roofed in, and there were little windows in it. It was beautifully clean and as tidy as possible. There were some lockers and boxes, and there was a table, and there was a Dutch clock, and there was a chest of drawers, and there was a tea-tray with a painting on it, and the tray was kept from tumbling down, by a Bible; and the tray, if it *had*[2] tumbled down, would have smashed a quantity of cups and saucers and a teapot that were grouped around the book. On the walls were coloured pictures of Abraham in red going to sacrifice Isaac in blue; and of Daniel in yellow being cast into a den of green lions.[3] Mr. Peggotty, as honest a seafaring man as ever breathed, dealt in lobsters, crabs and crawfish. *

As in my childhood, so in these days when I was a young man, Mr. Peggotty's household consisted of *his orphan nephew Ham Peggotty, a young shipwright; his adopted niece Little Emily, once my small sweetheart, now a beautiful young woman; and Mrs. Gummidge.*[4] All three had been maintained at Mr. Peggotty's sole charge for years and years; and Mrs. Gummidge was the widow of his partner in a boat, who had died poor. * She was very grateful, but she certainly would have been more agreeable[5] if she had not constantly complained, as she sat in the most comfortable corner by the fireside, that she was a 'lone lorn creetur and everythink went contrairy with her.'

Towards this old boat, I walked one memorable night, with *my former*

[1] *Berg* has the heading INTRODUCTION TO DAVID COPPERFIELD. Dickens did not delete this but wrote above it 'In all, Six chapters'. At the top of the page is a (rejected?) alternative opening phrase: 'I had known the odd dwelling house inhabited by Mr. Peggotty'. He began the reading 'Quickly' (Wright).

[2] *had* printed italic in *Berg* here and in the corresponding passage in the original Chapter I; it was not italic in the novel.

[3] Two passages in the remainder of this paragraph were deleted in *Berg* but had been printed in *T & F*: the first of a number of such late deletions.

[4] Small capitals in *Berg* here, because doubly underlined in the corresponding passage in the original Chapter I.

[5] *T & F* here includes some words deleted in *Berg*: '... more agreeable company in a small habitation ...'

schoolfellow and present dear friend, Steerforth;[1] Steerforth, half a dozen years older than I; brilliant, handsome, easy, winning; whom I admired with my whole heart; for whom I entertained the most romantic feelings of fidelity and friendship. He had come down with me from London, and had entered with the greatest ardour into my scheme of visiting the old simple place and the old simple people.

There was no moon; and as he and I walked on the dark wintry sands towards the old boat, the wind sighed mournfully.

'This is a wild place, Steerforth, is it not?'

'Dismal enough in the dark, and the sea has a cry in it, as if it were hungry for us. Is that the boat, where I see a light yonder?'

'That's the boat.'

We said no more as we approached the light, but made softly for the door. I laid my hand upon the latch; and whispering Steerforth to keep close to me, went in, † and I was in the midst of the astonished family, whom I had not seen from my childhood, face to face with Mr. Peggotty, and holding out my hand to him, when Ham shouted:

'Mas'r Davy! It's Mas'r Davy!'

In a moment we were all shaking hands with one another, and asking one another how we did, and telling one another how glad we were to meet, and all talking at once. Mr. Peggotty was so overjoyed to see me, and to see my friend, that he did not know what to say or do, but kept over and over again shaking hands with me, and then with Steerforth, and then with me, and then ruffling his shaggy hair all over his head, and then laughing with such glee and triumph, that it was a treat to see him.

'Why, that you two gentl'men—gentl'men growed—should come to this here roof to-night, of all nights in my life, is such a merry-go-rounder as never happened afore, I do rightly believe![2] Em'ly, my darling, come here! Come here, my little witch! *Theer's Mas'r Davy's friend, my dear! Theer's the gentl'man as you've heerd on, Em'ly.* He comes to see you, along with Mas'r Davy, on the brightest night of your uncle's life as ever was or will be, horroar for it!' Then he let her go;[3] and as she ran into her little chamber, looked round upon us, quite hot and out of breath with his uncommon satisfaction.

'If you two gentl'men—gentl'men growed now, and such gentl'men—don't ex-cuse me for being in a state of mind, when you understand

[1] Doubly underlined. The description of Steerforth does not appear in the novel.

[2] For Mr. Peggotty's voice, Dickens did not use 'the regular Yarmouth dialect, as I could not make it sufficiently intelligible in so large a place' as the halls where he performed (*N*, iii. 764).

[3] In the Reading text, as now printed, Mr. Peggotty had never had hold of her (he had, however, in a deleted passage 'put one of his large hands on each side of his niece's face'). But Dickens, as Peggotty, had in gesture chucked her under his arm 'just as if she were there *to* be chucked' (Field, p. 38).

matters, I'll arks your pardon. Em'ly, my dear!—She knows I'm agoing to tell, and has made off.[1] This here little Em'ly, sir,' to Steerforth, '—her as you see a blushing here just now—this here little Em'ly of ours, has been, in our house, sir, what I suppose (I'm a ignorant man, but that's my belief) no one but a little bright-eyed creetur *can*[2] be in a house. She ain't my child; I never had one; but I couldn't love her more, if she was fifty times my child. You understand! I couldn't do it!'

'I quite understand.'

'I know you do, sir, and thank'ee. † Well, Sir, there was a certain person as had know'd our Em'ly, from the time when her father was drownded; as had seen her constant when a babby, when a young gal, when a woman. Not much of a person to look at, he warn't—something o' my own build—rough—a good deal o' the sou'wester in him—wery salt—but, on the whole, a honest sort of a chap too, with his art in the right place.'

I had never seen Ham grin to anything like the extent to which he sat grinning at us now.

'What does this here blessed tarpaulin go and do, but he loses that there art of his to our little Em'ly. He follers her about, he makes hisself a sort o' servant to her, he loses in a great measure his relish for his wittles, and in the long run he makes it clear to me wot's amiss. † Well! I counsels him to speak to Em'ly. He's big enough, but he's bashfuller than a little un, and he says to me he doen't like. So *I* speak.[3] "What! *Him!*" says Em'ly. "*Him* that I've know'd so intimate so many year, and like so much! Oh, Uncle! I never can have *him*. He's such a good fellow!" I gives her a kiss, and I says no more to her than "My dear, you're right to speak out, you're to choose for yourself, you're as free as a little bird." Then I aways to him, and I says, "I wish it could have been so, but it can't. But you can both be as you was, and wot I say to you is, Be as you was with her, like a man." He says to me, a shaking of my hand, "I will!" he says. And he was—honourable, trew, and manful—going on for two year.

'All of a sudden, one evening—as it might be to-night—comes little Em'ly from her work, and him with her! There ain't so much in *that*, you'll say. No, sure, because he takes care on her, like a brother, arter dark, and indeed afore dark, and at all times. But this heer tarpaulin chap, he takes hold of her hand, and he cries out to me, joyful, "Lookee here! This is to be my little wife!" And she says, half bold and half shy, and half a laughing and half a crying, "Yes, uncle! If you please."—If I please! Lord, as if I should do anythink else!—"If you please,"

[1] Marginal stage-direction *Low*.

[2] *can* italic in *Berg* and in the novel.

[3] *I*, and all the other words italic in the rest of this speech, italic in *Berg* and in the novel.

she says, "I am steadier now, and I have thought better of it, and I'll be as good a little wife as I can to him, for he's a dear good fellow!" Then Missis Gummidge, she claps her hands like a play, and you come in. There! The murder's out! You come in! It took place this here present hour; and here's the man as'll marry her, the minute she's out of her time at the needlework.'

Ham staggered, as well he might, under the blow Mr. Peggotty dealt him, as a mark of confidence and friendship; but feeling called upon to say something to us, he stammered:[1]

'She warn't no higher than you was, Mas'r Davy—when you first come heer—when I thought what she'd grow up to be. I see her grow up—gentl'men—like a flower. I'd lay down my life for her—Mas'r Davy—Oh! most content and cheerful! There ain't a gentl'man in all the land—nor yet a sailing upon all the sea—that can love his lady more than I love her, though there's many a common man—as could say better—what he meant.'

I thought it affecting to see such a sturdy fellow trembling in the strength of what he felt for the pretty little creature who had won his heart. I thought the simple confidence reposed in us by Mr. Peggotty and by himself, was touching. I was affected by the story altogether. I was filled with pleasure; but at first, with an indescribably sensitive pleasure, that a very little would have changed to pain.

Therefore, if it had depended upon *me* to touch the prevailing chord among them with any skill, I should have made a poor hand of it. But it depended upon Steerforth; and he did it with such address, that in a few minutes we were all as easy as possible.

'Mr. Peggotty,' he said, 'you are a thoroughly good fellow, and deserve to be as happy as you are to-night. My hand upon it! Ham, I give you joy, my boy. My hand upon that, too![2] Davy, stir the fire, and make it a brisk one! And Mr. Peggotty, unless you can induce your gentle niece to come back, I shall go. Any gap at your fireside on such a night—such a gap least of all—I wouldn't make, for the wealth of the Indies!'

So, Mr. Peggotty went to fetch little Em'ly. At first little Em'ly didn't

[1] 'Feelingly' (Wright).

[2] A famous moment in the Reading: Dickens modified Steerforth's tone of voice, and hand-grip, as he 'turned' from Peggotty to Ham. It gave Dickens great satisfaction when, during a reading of *Copperfield* in Paris, a Frenchman in the front row 'suddenly exclaimed to himself, under his breath, "Ah-h!"—having instantly caught the situation!' (Kent, p. 124). According to another account, 'When I was impersonating Steerforth ... and gave that peculiar grip of the hand to Emily's lover, the French audience burst into cheers and rounds of applause' (J. T. Fields, *Yesterdays with Authors* (1872), p. 241). Maybe Dickens later sought too hard for further 'Ah-h's or cheers; certainly Kate Field thought he over-dramatized this moment—'for had Steerforth exhibited the hatred of Ham that darkens Dickens's face, it could not have passed unnoticed' (p. 39).

like to come, and then Ham went. Presently they brought her to the fireside, very much confused, and very shy,—but she soon became more assured when she found how Steerforth spoke to her; how skilfully he avoided anything that would embarrass her; how he talked to Mr. Peggotty of boats, and ships, and tides, and fish; how delighted he was with that boat and all belonging to it; how lightly and easily he carried on, until he brought us, by degrees, into a charmed circle. †

But he set up no monopoly of the conversation.[1] He was silent and attentive when little Emily talked across the fire to me of our old childish wanderings upon the beach, to pick up shells and pebbles; he was very silent and attentive, when I asked her if she recollected how I used to love her, and how we used to walk about that dim old flat, hours and hours, and how the days sported by us as if Time himself had not grown up then, but were a child like ourselves, and always at play.[2] She sat all the evening, in her old little corner by the fire—Ham beside her. *I could not satisfy myself whether it was in her little tormenting way, or in a maidenly reserve before us, that she kept quite close to the wall, and away from Ham; but I observed that she did so, all the evening.*

As I remember, it was almost midnight when we took our leave. We had had some biscuit and dried fish for supper, and Steerforth had produced from his pocket a flask of Hollands. We parted merrily; and as they all stood crowded round the door to light us on our road, I saw the sweet blue eyes of little Em'ly peeping after us, from behind Ham, and heard her soft voice calling to us to be careful how we went.

'A most engaging little Beauty!' *said Steerforth, taking my arm.*[3] 'Well! It's a quaint place, and they are quaint company; and it's quite a new sensation to mix with them.'

'How fortunate we are, too, Steerforth, to have arrived to witness their happiness in that intended marriage! I never saw people so happy. How delightful to see it!'

'Yes—that's rather a chuckle-headed fellow for the girl. Isn't he?'

I felt a shock in this cold reply. But turning quickly upon him, and seeing a laugh in his eyes, I answered:

'Ah, Steerforth! It's well for you to joke about the poor! But when I see how perfectly you understand them, and how you can enter into happiness like this plain fisherman's, I know there is not a joy, or sorrow, or any emotion, of such people, that can be indifferent to you. And I admire and love you for it, Steerforth, twenty times the more!' *

To my surprise, he suddenly said, with nothing, that I could see, to lead to it:[4]

'Daisy, I wish to God I had had a judicious father these last twenty

[1] 'Leaning on table' (Wright). [2] From ch. 3 (p. 37). [3] 'Rough' (Wright).
[4] Indeed not: the text jumps here from ch. 21 to ch. 22.

years! You know my mother has always doted on me and spoilt me. I wish with all my soul I had been better guided! I wish with all my soul, I could guide myself better!'

There was a passionate dejection in his manner that quite amazed me. He was more unlike himself than I could have supposed possible.

'It would be better to be this poor Peggotty, or his lout of a nephew,[1] than be myself, twenty times richer and twenty times wiser, and be the torment to myself that I have been in that Devil's bark of a boat within the last half-hour.'

I was so confounded by the change in him that at first I could only regard him in silence as he walked at my side. At length I asked him to tell me what had happened to cross him so unusually.

'Tut, it's nothing—nothing, Davy! I must have had a nightmare, I think. What old women call the horrors, have been creeping over me from head to foot. I have been afraid of myself.'

'You are afraid of nothing else, I think.'

'Perhaps not, and yet may have enough to be afraid of, too. Well! so it goes by![2] * Daisy—for though that's not the name your godfathers and godmothers gave you, you're such a fresh fellow that it's the name I best like to call you by—and I wish, I wish, I wish, you could give it to me!'

'Why, so I can, if I choose.'

'Daisy, if anything should ever happen to separate us, you must think of me at my best, old boy. Come! let us make that bargain. Think of me at my best, if circumstances should ever part us!'

'You have no best to me, Steerforth, and no worst. You are always equally loved and cherished in my heart.' *

I was up, to go away alone, next morning with the dawn, and, having dressed as quietly as I could, looked into his room. He was fast asleep; lying, easily, with his head upon his arm, as I had often seen him lie at school.

The time came in its season, and that was very soon, when I almost wondered that nothing troubled his repose, as I looked at him then.[3] *But he slept—let me think of him so again—as I had often seen him sleep at school; and thus, in this silent hour I left him.*

—Never more, O God forgive you, Steerforth! to touch that passive hand in love and friendship. Never, never, more!

[1] 'Looking up occasionally with a sneer' (Wright). [2] Jump to ch. 29.

[3] This paragraph 'Solemnly, slowly, measuredly, and with feeling' (Wright). Dickens's rendering of 'Never more, . . . Steerforth! . . .' was widely admired. Kate Field doubted whether any actor could equal his rendering of 'the solemn yet tender sorrow' of this exclamation (p. 39). 'Sigh' on final words (Wright).

CHAPTER II[1]

SOME months elapsed, before I again found myself down in that part of the country, and approaching the old boat by night.[2]

It was a dark evening, and rain was beginning to fall, when I came within sight of Mr. Peggotty's house, and of the light within it shining through the window. A little floundering across the sand, which was heavy, brought me to the door, and I went in. I was bidden to a little supper; Emily was to be married to Ham that day fortnight, and this was the last time I was to see her in her maiden life.

It looked very comfortable, indeed. Mr. Peggotty had smoked his evening pipe, and there were preparations for supper by-and-by. The fire was bright, the ashes were thrown up, *the locker was ready for little Emily in her old place*. Mrs. Gummidge appeared to be fretting a little, in her own corner: and consequently looked quite natural.

'You're first of the lot, Mas'r Davy! Sit ye down, sir. It ain't o' no use saying welcome to you, but you're welcome, kind and hearty.' *

Here Mrs. Gummidge groaned.

'Cheer up, cheer up, Mrs. Gummidge!' said Mr. Peggotty. *

'No, no, Dan'l. It ain't o' no use telling *me* to cheer up, when everythink goes contrairy with me. Nothink's nat'ral to me but to be lone and lorn.'

After looking at Mrs. Gummidge for some moments, with great sympathy, Mr. Peggotty glanced at the Dutch clock, rose, snuffed the candle, and put it in the window.

'Theer! Theer we are, Missis Gummidge!'[3] *Mrs. Gummidge slightly groaned again.* 'Theer we are, Mrs. Gummidge, lighted up, accordin' to custom! You're a wonderin' what that's fur, sir! Well, it's fur our little Em'ly. You see, the path ain't over light or cheerful arter dark; and when I'm here at the hour as she's a comin' home from her needlework down-town, I puts the light in the winder. That, you see, meets two objects. She says to herself, says Em'ly, "Theer's home!" she says. And likeways, says Em'ly, "My uncle's theer!" Fur if I ain't theer, I never have no light showed. You may say this is like a Babby, sir. Well, I doen't know but what I *am* a babby in regard o' Em'ly. Not to look

[1] In *Berg*, CHAPTER THE FIRST is amended to SECOND. Similar changes are made throughout, and will not further be noted. The opening pages of this Chapter are deleted, having been transferred to the new beginning; a new opening paragraph is written in.

[2] The paragraph continued: 'I remember the occasion well. Events of later date have floated from me to the shore where all forgotten things will re-appear; but *this* stands like a high rock in the Ocean.' These sentences (drawn from ch. 9, Oxford Illustrated edn., p. 131) were later deleted.

[3] 'Jolly' (Wright). Later in the paragraph, near 'That, you see, meets two objects', Wright has: 'Facial expression. Elevating eyebrows'.

at, but to—to consider on, you know. *I* doen't care, bless you![1] Now I tell you. When I go a looking and looking about that theer pritty house of our Em'ly's, all got ready for her to be married, if I doen't feel as if the littlest things was her, a'most. I takes 'em up, and I puts 'em down, and I touches of 'em as delicate as if they was our Em'ly. So 't is with her little bonnets and that. I couldn't see one on 'em rough used a purpose—not fur the whole wureld.

'It's my opinion, you see, as this is along of my havin' played with Em'ly so much when she was a child, and havin' made believe as we was Turks, and French, and sharks, and every wariety of forinners—bless you, yes; and lions and whales, and I don't know what all!—when she warn't no higher than my knee. I've got into the way on it, you know. Why, this here candle, now! *I*[2] know wery well that arter she's married and gone, I shall put that candle theer, just the same as now, and sit afore the fire, pretending I'm expecting of her, like as I'm a doing now. Why, at the present minute, when I see the candle sparkle up, I says to myself, "She's a looking at it! Em'ly's a coming!" Right too, fur here she is!'

No; it was only Ham. The night should have turned more wet since I came in, for he had a large sou'wester hat on, slouched over his face.

'Where's Em'ly?'

Ham made a movement, as if she were outside. Mr. Peggotty took the light from the window, trimmed it, put it on the table, and was stirring the fire, when Ham, who had not moved, said:

'Mas'r Davy, will you come out a minute, and see what Em'ly and me has got to show you?'[3]

As I passed him, I saw, to my astonishment and fright, that he was deadly pale. He closed the door upon us. Only upon us two.

'Ham! What's the matter?'

'My love, Mas'r Davy—the pride and hope of my art—her that I'd have died for, and would die for now—she's gone!'

'Gone!'

'Em'ly's run away![4] You're a scholar, and know what's right and best. What am I to say, in-doors? How am I ever to break it to him, Mas'r Davy?'

I saw the door move, and tried to hold the latch, to gain a moment's time.[5] It was too late. Mr. Peggotty thrust forth his face; and never could I forget the change that came upon it when he saw us, if I were to live five hundred years.

[1] *I* italic in *Berg* and in the novel.

[2] *I* italic in *Berg* and in the novel.

[3] 'Breathlessly and slow' (Wright).

[4] 'D D D' (Wright)—i.e. very Dramatic. Dickens had, however, deleted from this episode Ham's weeping; Kent (p. 127) regretted this.

[5] 'Fast' (Wright).

I remember a great wail and cry, and the women hanging about him, and we all standing in the room; I with an open letter in my hand, which Ham had given me; Mr. Peggotty, with his vest torn open, his hair wild, his face and lips white, and blood trickling down his bosom (it had sprung from his mouth, I think).

'Read it, sir; slow, please. I doen't know as I can understand.'

In the midst of the silence of death, I read thus, from the blotted letter Ham had given me. In Em'ly's hand—addressed to himself:

'"When you, who love me so much better than I ever have deserved, even when my mind was innocent, see this, I shall be far away. When I leave my dear home—my dear home—oh, my dear home!—in the morning,"'—the letter bore date on the previous night: '"—it will be never to come back, unless he brings me back a lady. This will be found at night, many hours after, instead of me. For mercy's sake, tell uncle that I never loved him half so dear as now. Oh, don't remember you and I were ever to be married—but try to think *as if I died when I was little*,[1] *and was buried somewhere*. Pray Heaven *that I am going away from*, have compassion on my uncle! Be his comfort. Love some good girl, that will be what I was once to uncle, and that will be true to you, and worthy of you, *and know no shame but me*. God bless all! If he don't bring me back a lady, and I don't pray for my own self, I'll pray for all. My parting love to uncle. My last tears, and my last thanks, for uncle!"'

That was all.

He stood, long after I had ceased to read, still looking at me. Slowly, at last, he moved his eyes from my face, and cast them round the room.

'Who's the man? I want to know his name.'

Ham glanced at me, and suddenly I felt a shock.

'Mas'r Davy! Go out a bit, and let me tell him what I must. You doen't ought to hear it, sir.'

I sank down in a chair, and tried to utter some reply; but my tongue was fettered, and my sight was weak. For I felt that the man was my friend—the friend I had unhappily introduced there—Steerforth, my old schoolfellow and my friend.

'I want to know his name!' †

'Mas'r Davy's frend. He's the man. Mas'r Davy, it ain't no fault of yourn—and I am far from laying of it to you—but it is your friend Steerforth, and he's a damned villain!'

Mr. Peggotty moved no more, until he seemed to wake all at once, and pulled down his rough coat from its peg in a corner.

'Bear a hand with this! I'm struck of a heap, and can't do it. Bear a hand, and help me. Well! Now give me that theer hat!'

Ham asked him whither he was going?

[1] *T&F* reads 'very little'. Dickens read this letter 'As if Emily was speaking,' 'Slow and pausing' (Wright). Thackeray regarded it as 'a masterpiece', his daughter Anne recorded.

'I'm a going to seek my niece. I'm a going to seek my Em'ly. I'm a going, first, to stave in that theer boat as he gave me, and sink it where I would have drownded *him*,[1] as I'm a livin' soul, if I had had one thought of what was in him! As he sat afore me, in that boat, face to face, strike me down dead, but I'd have drownded him, and thought it right!—I'm a going fur to seek my niece.'[2]

'Where?'

'Anywhere! I'm a going to seek my niece through the wureld. I'm a going to find my poor niece in her shame, and bring her back wi' my comfort and forgiveness. No one stop me! I tell you I'm a going to seek my niece! I'm a going to seek her fur and wide!'

Mrs. Gummidge came between them, in a fit of crying. 'No, no, Dan'l, not as you are now. Seek her in a little while, my lone lorn Dan'l, and that'll be but right; but not as you are now. Sit ye down, and give me your forgiveness for having ever been a worrit to you, Dan'l—what have *my*[3] contrairies ever been to this!—and let us speak a word about them times when she was first a orphan, and when Ham was too, and when I was a poor widder woman, and you took me in.[4] It'll soften your poor heart, Dan'l, and you'll bear your sorrow better; for you know the promise, Dan'l, "As you have done it unto one of the least of these, you have done it unto me;" and that can never fail under this roof, that's been our shelter for so many, many year!'

He was quite passive now; and when I heard him crying, the impulse that had been upon me to go down upon my knees, and curse Steerforth, yielded to a better feeling. My overcharged heart found the same relief as his, and I cried too.

CHAPTER III

AT this period of my life I lived in my top set of chambers in Buckingham Street, Strand, London, and was over head and ears in love with Dora. I lived principally on Dora and coffee. My appetite languished and I was glad of it, for I felt as though it would have been an act of perfidy towards Dora to have a natural relish for my dinner. * I bought four sumptuous waistcoats—not for myself; *I*[5] had no pride in them—for Dora. I took to wearing straw-coloured kid gloves in the streets. I laid the foundations of all the corns I have ever had. *If the boots I wore at that period could only be produced and compared with the natural size of*

[1] *him* italic in *Berg* and in the novel.

[2] It was the rendering of Peggotty here, as his mind wandered in rage, grief and bewilderment, that reminded one critic of King Lear (*Manchester Examiner*, 19 October 1868; cf. Field, p. 41).

[3] *my* italic in *Berg* and in the novel.

[4] 'Both hands up. Monotone' (Wright).

[5] *I* italic in *Berg* and in the novel.

my feet, they would show in a most affecting manner what the state of my heart was.[1] *

Mrs. Crupp, the housekeeper of my chambers, must have been a woman of penetration; for, when this attachment was but a few weeks old, she found it out. She came up to me one evening when I was very low, to ask (she being afflicted with spasms) if I could oblige her with a little tincture of cardamums, mixed with rhubarb and flavoured with seven drops of the essence of cloves—or, if I had not such a thing by me—with a little brandy. As I had never even heard of the first remedy, and always had the second in the closet, I gave Mrs. Crupp a glass of the second; which (that I might have no suspicion of its being devoted to any improper use) she began to take immediately.

'Cheer up, sir,' said Mrs. Crupp. 'Excuse me. I know what it is, sir. There's a lady in the case.'

'Mrs. Crupp?'

'Oh, bless you! Keep a good heart, sir! Never say die, sir! If she don't smile upon you, there's a many as will. You're a young gentleman to *be*[2] smiled on, Mr. Copperfull, and you must learn your walue, sir.'

Mrs. Crupp always called me Mr. Copperfull: firstly, no doubt, because it was not my name; and secondly, I am inclined to think, in some indistinct association with a washing-day.

'What makes you suppose there is any young lady in the case, Mrs. Crupp?'

'Mr. Copperfull, I'm a mother myself. * Your boots and your waist is equally too small, and you don't eat enough, sir, nor yet drink. Sir, I have laundressed other young gentlemen besides you. † It was but the gentleman which died here before yourself, that fell in love—with a barmaid—and had his waistcoats took in directly, though much swelled by drinking.'

'Mrs. Crupp, I must beg you not to connect the young lady in my case with a barmaid, or anything of that sort, if you please.'

'Mr. Copperfull, I'm a mother myself, and not likely. I ask your pardon, sir, if I intrude. I should never wish to intrude where I were not welcome. But you are a young gentleman, Mr. Copperfull, and my adwice to you is, to cheer up, sir, to keep a good heart, and to know

[1] With a 'falling inflection' to the end of the sentence (Wright). The effect was helped by the phrase 'in a most affecting manner' having been moved from the end of the sentence, where it occurs in the novel, to its present place.—The present editor, having often performed this item, might draw attention to this, as an instance of Dickens's professional skill, because the alteration makes all the difference. David's *heart*, thus brought to the end of the paragraph, and played off against his emotionally ignominious *feet*, can hardly fail to make an audience laugh. No laugh here, indeed, presages a tough evening ahead.

[2] *be* italic in *Berg* and in the novel. Against this dialogue with Mrs. Crupp, Wright has: 'No action. Look about from place to place.'

your own walue. If you was to take to something, sir; if you was to take to skittles, now, which is healthy, you might find it divert your mind, and do you good.'

I turned it off and changed the subject by informing Mrs. Crupp that I wished to entertain at dinner next day, my esteemed friends Traddles, and Mr. and Mrs. Micawber. And I took the liberty of suggesting a pair of soles, a small leg of mutton, and a pigeon pie. Mrs. Crupp broke out into rebellion on my first bashful hint in reference to *her* cooking the fish and joint. But, in the end, a compromise was effected; and Mrs. Crupp consented to achieve this feat, on condition that I dined from home for a fortnight afterwards. * †

Having laid in the materials for a bowl of punch, to be compounded by Mr. Micawber; having provided a bottle of lavender-water, two wax candles, a paper of mixed pins, and a pin-cushion, to assist Mrs. Micawber in her toilette, at my dressing-table; having also caused the fire in my bed-room to be lighted for Mrs. Micawber's convenience; and having laid the cloth with my own hands; I awaited the result with composure.

At the appointed time, my three visitors arrived together. Mr. Micawber with more shirt-collar than usual, and a new ribbon to his eye-glass; Mrs. Micawber with her cap in a parcel; Traddles carrying the parcel, and supporting Mrs. Micawber on his arm. They were all delighted with my residence. When I conducted Mrs. Micawber to my dressing-table, and she saw the scale on which it was prepared for her, she was in such raptures, that she called Mr. Micawber to come in and look.

'My dear Copperfield,' said Mr. Micawber, 'this is luxurious.[1] This is a way of life which reminds me of the period when I was myself in a state of celibacy. * I am at present established on what may be designated as a small and unassuming scale; but, you are aware that I have, in the course of my career, surmounted difficulties, and conquered obstacles. You are no stranger to the fact, that there have been periods of my life, when it has been requisite that I should pause, until certain expected events should turn up—when it has been necessary that I should fall back, before making what I trust I shall not be accused of presumption in terming—a spring. The present is one of those momentous stages in the life of man. You find me, fallen back, *for*[2] a spring; and I have every reason to believe that a vigorous leap will shortly be the result.'

I informed Mr. Micawber that I relied upon him for a bowl of punch, and led him to the lemons. I never saw a man so thoroughly enjoy himself, as he stirred, and mixed, and tasted, and looked as if he were

[1] Micawber 'Shaking head', 'Slight cough. Swinging eyeglass' (Wright). Kate Field notes also that Dickens, as Micawber, tipped 'backward and forward, first on his heels and then on his toes' and that he savoured his words, e.g. 'lux-u-rious' (p. 43).

[2] *for* italic in *Berg* and in the novel.

making, not mere punch, but a fortune for his family down to the latest posterity. As to Mrs. Micawber, I don't know whether it was the effect of the cap, or the lavender-water, or the pins, or the fire, or the wax-candles, but she came out of my room, comparatively speaking, lovely.[1]

I suppose—I never ventured to inquire, but I suppose—that Mrs. Crupp, after frying the soles, was taken ill. Because we broke down at that point. The leg of mutton came up, very red inside, and very pale outside: besides having a foreign substance of a gritty nature sprinkled over it, as if it had had a fall into ashes. But we were not in a condition to judge of this fact from the appearance of the gravy, forasmuch as it had been all dropped on the stairs. The pigeon-pie was not bad, but it was a delusive pie[2]: the crust being like a disappointing phrenological head: *full of lumps and bumps, with nothing particular underneath.* In short, the banquet was such a failure that I should have been quite unhappy—about the failure, I mean, for I was always unhappy about Dora—if I had not been relieved by the great good-humour of my company.

'My dear friend Copperfield,' said Mr. Micawber, 'accidents will occur in the best-regulated families; and especially in families not regulated by that pervading influence which sanctifies while it enhances the—a—I would say, in short, by the influence of Woman in the lofty character of Wife. If you will allow me to take the liberty of remarking that there are few comestibles better, in their way, than a Devil, and that I believe, with a little division of labour, we could accomplish a good one if the young person in attendance could produce a gridiron, I would put it to you, that this little misfortune may be easily repaired.'

There *was* a gridiron in the pantry, on which my morning rasher of bacon was cooked. We had it out, in a twinkling; Traddles cut the mutton into slices; Mr. Micawber covered them with pepper, mustard, salt, and cayenne;[3] I put them on the gridiron, turned them with a fork, and took them off, under Mr. Micawber's direction; and Mrs. Micawber heated some mushroom ketchup in a little saucepan. † Under these circumstances, my appetite came back miraculously. I am ashamed to confess it, but I really believe I forgot Dora for a little while. *

'Punch, my dear Copperfield,' said Mr. Micawber, tasting it as soon as dinner was done,[4] 'like time and tide, waits for no man. Ah! it is at the present moment in high flavour. My love, will you give me your opinion?'

[1] Dickens's audiences roared with laughter when, as David, he prolonged the final word, 'l-l-lovely!' (Kent, p. 129).

[2] Dickens paused before the adjective 'delusive' (Wright), and made much of it (Field, p. 46).

[3] 'Motions over table' (Wright); *was* (above) doubly underlined.

[4] Marginal stage-direction *Tasting*. Littimer's entrance here is omitted.

Mrs. Micawber pronounced it excellent. †

'As we are quite confidential here, Mr. Copperfield,' said Mrs. Micawber sipping her punch,[1] '(Mr. Traddles being a part of our domesticity), I should much like to have your opinion on Mr. Micawber's prospects. I have consulted branches of my family on the course most expedient for Mr. Micawber to take, and it was, that he should immediately turn his attention to coals.'

'To what, ma'am?'

'To coals. To the coal trade. Mr. Micawber was induced to think, on inquiry, that there might be an opening for a man of his talent in the Medway Coal Trade. Then, as Mr. Micawber very properly said, the first step to be taken clearly was, to go and *see*[2] the Medway. Which we went and saw. I say "we," Mr. Copperfield; for I never will desert Mr. Micawber.[3] I am a wife and mother, and I never will desert Mr. Micawber.'

Traddles and I murmured our admiration.

'That,' said Mrs. Micawber, 'that, at least, is *my*[4] view, my dear Mr. Copperfield and Mr. Traddles, of the obligation which I took upon myself when I repeated the irrevocable words "I Emma, take thee, Wilkins." I read the service over with a flat-candle, on the previous night, and the conclusion I derived from it was that I never could or would desert Mr. Micawber.'

'My dear,' said Mr. Micawber, *a little impatiently*, 'I am not conscious that you are expected to do anything of the sort.'

'We went,' repeated Mrs. Micawber, 'and saw the Medway. My opinion of the coal trade on that river, was, that it might require talent, but that it certainly requires capital. Talent, Mr. Micawber has; capital, Mr. Micawber has not. We saw, I think, the greater part of the Medway; and that was my individual conclusion.[5] My family were then of opinion that Mr. Micawber should turn his attention to corn—on commission. But corn, as I have repeatedly said to Mr. Micawber, may be gentlemanly, but it is not remunerative. Commission to the extent of two and ninepence in a fortnight cannot, however limited our ideas, be considered remunerative.'

We were all agreed upon that.

'Then,' said Mrs. Micawber, *who prided herself on taking a clear view of things, and keeping Mr. Micawber straight by her woman's wisdom, when he might otherwise go a little crooked*, * 'then I naturally look round

[1] Marginal stage-direction *Sipping*. Wright notes further sips, during the ensuing speeches. The ensuing Coal Trade passage is from ch. 17.

[2] *see* italic in *Berg* and in the novel.

[3] 'Affectedly' (Wright). This 'never desert . . .' passage is from ch. 36.

[4] *my* italic in *Berg*, but not in the novel.

[5] 'Very affectedly' (Wright). Back to ch. 28, for corn and banking.

the world, and say, "What is there in which a person of Mr. Micawber's talent is likely to succeed?" † I may have a conviction that Mr. Micawber's manners peculiarly qualify him for the Banking business. I may argue within myself, that if *I*[1] had a deposit at a banking-house, the manners of Mr. Micawber, as representing that banking-house, would inspire confidence, and extend the connexion. But if the various banking-houses refuse to avail themselves of Mr. Micawber's abilities, or receive the offer of them with contumely, what is the use of dwelling upon *that*[2] idea? None. As to originating a banking-business, I may know that there are members of my family who, if they chose to place their money in Mr. Micawber's hands, might found an establishment of that description. But if they do *not* choose to place their money in Mr. Micawber's hands—which they don't—what is the use of that?[3] Again I contend that we are no farther advanced than we were before.'

I shook my head, and said, 'Not a bit.' Traddles also shook his head, and said, 'Not a bit.'

'What do I deduce from this?' *Mrs Micawber went on to say, still with the same air of putting a case lucidly*. 'What is the conclusion, my dear Mr. Copperfield, to which I am irresistibly brought? Am I wrong in saying, it is clear that we must live?'

I answered, 'Not at all!'[4] and Traddles answered, 'Not at all!' and I found myself afterwards sagely adding, alone, that a person must either live or die.

'Just so,' returned Mrs. Micawber. 'It is precisely that. † And here is Mr. Micawber without any suitable position or employment. Where does that responsibility rest? Clearly on society. Then I would make a fact so disgraceful known, and boldly challenge society to set it right. It appears to me, my dear Mr. Copperfield, that what Mr. Micawber has to do is to throw down the gauntlet to society, and say, in effect, "Show me who will take that up. Let the party immediately step forward." It appears to me, that what Mr. Micawber has to do, is to advertise in all the papers; to describe himself plainly as so and so, with such and such qualifications, and to put it thus: "*Now*[5] employ me, on remunerative terms, and address, post paid, to *W. M.*, Post Office, Camden Town." † For this purpose, I think Mr. Micawber ought to raise a certain sum of money—on a bill. If no member of my family is possessed of sufficient natural feeling to negotiate that bill, then, my opinion is, that Mr. Micawber should go into the City, should take that bill into the Money Market, and should dispose of it for what he can get.'

[1] *I* italic in *Berg*, but not in the novel.

[2] *that*, like *not* (below), italic in *Berg* and in the novel.

[3] 'Elevate chin' (Wright).

[4] 'Quickly' (Wright). Traddles, writes Kate Field (p. 43), sprang to life with his 'Not at all!' as a distinct character 'with a propensity to eat his own fingers'.

[5] *Now*, like *W.M.* (below), italic in *Berg* and in the novel.

I felt, but I am sure I don't know why, that this was highly self-denying and devoted in Mrs. Micawber, and I uttered a murmur to that effect. Traddles, who took his tone from me, did likewise, * and really I felt that she was a noble woman—the sort of woman who might have been a Roman matron, and done all manner of troublesome heroic public actions.[1]

In the fervour of this impression, I congratulated Mr. Micawber on the treasure he possessed. So did Traddles. Mr. Micawber extended his hand to each of us in succession, and then covered his face with his pocket-handkerchief—which I think had more snuff upon it than he was aware of. He then returned to the punch in the highest state of exhilaration. †

Mrs. Micawber made tea for us in a most agreeable manner; and after tea we discussed a variety of topics before the fire; and she was good enough to sing us (in a small, thin, flat voice, which I remembered to have considered, when I first knew her, the very table-beer of acoustics) the favourite ballads of 'The Dashing White Sergeant,' and 'Little Tafflin.' For both of these songs Mrs. Micawber had been famous when she lived at home with her papa and mamma. Mr. Micawber told us, that when he heard her sing the first one, on the first occasion of his seeing her beneath the parental roof, she had attracted his attention in an extraordinary degree; but that when it came to Little Tafflin, he had resolved to win that woman, or perish in the attempt.

It was between ten and eleven o'clock when Mrs. Micawber rose to replace her cap in the parcel, and to put on her bonnet. Mr. Micawber took the opportunity to slip a letter into my hand, with a whispered request that I would read it at my leisure. *I*[2] also took the opportunity of my holding a candle over the bannisters to light them down, when Mr. Micawber was going first, leading Mrs. Micawber, to detain Traddles for a moment on the top of the stairs.

'Traddles, Mr. Micawber don't mean any harm; but, if I were you, I wouldn't lend him anything.'

'My dear Copperfield, I haven't got anything to lend.'

'You have got a name, you know.'

'Oh! you call *that*[3] something to lend?'

'Certainly.'

'Oh! Yes, to be sure! I am very much obliged to you, Copperfield, but—I am afraid I have lent him that already.'

'For the bill that is to go into the Money Market?'

[1] A long deletion follows in *Berg*, with the next two pages being joined by stamp-edging: the text resumed at 'It was between ten and eleven o'clock'. Later, Dickens *stetted* some passages, and had to break the stamp-edging.

[2] *I* doubly underlined.

[3] *that* italic in *Berg* and in the novel.

'No. Not for that one. This is the first I have heard of that one. I have been thinking that he will most likely propose that one, on the way home. Mine's another.'

'I hope there will be nothing wrong about it.'

'I hope not. I should think not, though, because he told me, only the other day, that it was provided for. That was Mr. Micawber's expression, "Provided for."'

Mr. Micawber looking up at this juncture, I had only time to repeat my caution. Traddles thanked me, and descended. But I was much afraid, when I observed the good-natured manner in which he went down with Mrs. Micawber's cap in his hand, that he would be carried into the Money Market, neck and heels.[1]

I returned to my fireside, and[2] * read Mr. Micawber's letter, which was dated an hour and a half before dinner. I am not sure whether I have mentioned that, when Mr. Micawber was at any particularly desperate crisis, he used a sort of legal phraseology: which he seemed to think equivalent to winding up his affairs.

This was the letter.

'Sir—for I dare not say my dear Copperfield,

'It is expedient that I should inform you that the undersigned is Crushed. Some flickering efforts to spare you the premature knowledge of his calamitous position, you may observe in him this day; but hope has sunk beneath the horizon, and the undersigned is Crushed.

'The present communication is penned within the personal range (I cannot call it the society) of an individual, in a state closely bordering on intoxication, employed by a broker. That individual is in legal possession of the premises, under a distress for rent. His inventory includes, not only the chattels and effects of every description belonging to the undersigned, as yearly tenant of this habitation, but also those appertaining to Mr. Thomas Traddles, lodger, a member of the Honourable Society of the Inner Temple.

'If any drop of gloom were wanting in the overflowing cup, which is now "commended" (in the language of an immortal Writer) to the lips of the undersigned, it would be found in the fact, that a friendly acceptance granted to the undersigned, by the before-mentioned Mr. Thomas Traddles, for the sum of £23 4*s*. 9½*d*., is over due, and is NOT provided for. Also, in the fact, that the living responsibilities clinging to the undersigned will, in the course of nature, be increased by the sum of one more helpless victim; whose miserable appearance may be looked for—in round numbers—at the expiration of a period not exceeding six lunar months from the present date.

[1] Chapter III ended at this point, Dickens having deleted the rest. Later he *stetted* it.

[2] A long omission here—Steerforth's visit to David.

'After premising thus much, it would be a work of supererogation to add, that dust and ashes are for ever scattered

'On
'The
'Head
'Of
'WILKINS MICAWBER.'

CHAPTER IV

SELDOM did I wake at night, seldom did I look up at the moon or stars or watch the falling rain, or hear the wind, but I thought of the solitary figure of the good fisherman toiling on—poor Pilgrim!—and recalled his words, 'I'm a going to seek my niece. I'm a going to seek her fur and wide.'

Months passed, and he had been absent—no one knew where—the whole time.

It had been a bitter day in London, and a cutting north-east wind had blown. The wind had gone down with the light, and snow had come on. My shortest way home,—and I naturally took the shortest way on such a night—was through Saint Martin's Lane. * On the steps of the church, there was the figure of a man. And I stood face to face with Mr. Peggotty!

'Mas'r Davy! It do my art good to see you, sir. Well met, well met!'

'Well met, my dear old friend!'

'I had thowts o' coming to make inquiration for you, sir, to-night, but it was too late. I should have come early in the morning, sir, afore going away agen.'

'Again?'

'Yes, sir, I'm away to-morrow.'

In those days there was a side entrance to the stable-yard of the Golden Cross Inn. Two or three public-rooms opened out of the yard: and looking into one of them, and finding it empty, and a good fire burning, I took him in there. †

'I'll tell you, Mas'r Davy, wheer-all I've been, and what-all we've heerd.[1] I've been fur, and we've heerd little; but I'll tell you!'

As he sat thinking, there was a fine massive gravity in his face, which I did not venture to disturb.

'You see, sir, when she was a child, she used to talk to me a deal about the sea, and about them coasts where the sea got to be dark blue, and to lay a shining and a shining in the sun. When she was—lost, I

[1] 'Feelingly' (Wright)—repeated against Peggotty's next speech.

know'd in my mind, as he would take her to them countries.[1] I know'd in my mind, as he'd have told her wonders of 'em, and how she was to be a lady theer, and how he first got her to listen to him along o' sech like. I went across-channel to France, and landed theer, as if I'd fell down from the skies. I found out a English gentleman, as was in authority, and told him I was going to seek my niece. He got me them papers as I wanted fur to carry me through—I doen't rightly know how they're called—and he would have give me money, but that I was thankful to have no need on. I thank him kind, for all he done, I'm sure! I told him, best as I was able, what my gratitoode was, and went away through France, fur to seek my niece.'

'Alone, and on foot?'

'Mostly a-foot; sometimes in carts along with people going to market; sometimes in empty coaches. Many mile a day a-foot, and often with some poor soldier or another, travelling fur to see his friends. I couldn't talk to him, nor he to me; but we was company for one another, too, along the dusty roads. When I come to any town, I found the inn, and waited about the yard till some one came by (some one mostly did) as know'd English. Then I told how that I was on my way to seek my niece, and they told me what manner of gentlefolks was in the house, and I waited to see any as seemed like her, going in or out. When it warn't Em'ly, I went on agen. By little and little, when I come to a new village or that, among the poor people, I found they know'd about me. They would set me down at their cottage doors, and give me what-not fur to eat and drink, and show me where to sleep.[2] And many a woman, Mas'r Davy, as has had a daughter of about Em'ly's age, I've found a-waiting for me, at Our Saviour's Cross outside the village, fur to do me sim'lar kindnesses. Some has had daughters as was dead. And God only knows how good them mothers was to me!' *

I laid my trembling hand upon the hand he put before his face. 'Thankee, sir, doen't take no notice.'

'At last I come to the sea. It warn't hard, you may suppose, for a seafaring man like me to work his way over to Italy. When I got theer, I wandered on as I had done afore. I got news of her being seen among them Swiss mountains yonder. I made for them mountains, day and night. Ever so fur as I went, ever so fur them mountains seemed to shift *away from me*. But I come up with 'em, and I crossed 'em. I never doubted her. No! Not a bit! On'y let her see my face—on'y let her heer my voice—on'y let my stanning still afore her bring to her thoughts the home she had fled away from, and the child she had been—and if

[1] 'Slow, low, monotonous' (Wright). Against the next paragraph but one ('Mostly a-foot . . .') Wright again puts 'Monotonously': but, as others of his marginalia show, the term was not intended pejoratively.

[2] 'Hand over eyes' (Wright).

she had growed to be a royal lady, she'd have fell down at my feet! I know'd it well! I bought *a country dress* to put upon her. To put that dress upon her, and to cast off what she wore—to take her on my arm again, and wander towards home—to stop sometimes upon the road, and heal her bruised feet and her worse-bruised heart—was all I thowt of now. But, Mas'r Davy, it warn't to be—not yet! I was too late, and they was gone. Wheer, I couldn't learn. Some said heer, some said theer. I travelled heer, and I travelled theer, but I found no Em'ly, and I travelled home.'

'How long ago?'

'A matter o' fower days. I sighted the old boat arter dark, and I never could have thowt, I'm sure, that the old boat would have been so strange!'

From some pocket in his breast, he took out with a very careful hand, a small paper bundle containing two or three letters or little packets, which he laid upon the table.[1]

'The faithful creetur Mrs. Gummidge gave me these. This first one come afore I had been gone a week. A fifty pound Bank note, in a sheet of paper, directed to me, and put underneath the door in the night. She tried to hide her writing, but she couldn't hide it from Me! This one come to Missis Gummidge, two or three months ago. * Five pounds.'

It was untouched like the previous sum, and he refolded both. *

'Is that another letter in your hand?'

'It's money too, sir. Ten pound, you see. And wrote inside, "From a true friend." But the two first was put underneath the door, and this come by the post, day afore yesterday. I'm going to seek her at the post-mark.'

He showed it to me. It was a town on the Upper Rhine. He had found out, at Yarmouth, some foreign dealers who knew that country, and they had drawn him a rude map on paper, which he could very well understand.

I asked him how Ham was?

'He works as bold as a man can. He's never been heerd fur to complain. But my belief is ('twixt ourselves) as it has cut him deep. * Well! Having seen you to-night, Mas'r Davy (and that doos me good!), I shall away betimes to-morrow morning. You have seen what I've got heer;' putting his hand on where the little packet lay; 'all that troubles me is, to think that any harm might come to me, afore this money was give back. If I was to die, and it was lost, or stole, or elseways made away with, and it was never know'd by him but what I'd accepted of it, I believe the t'other wureld wouldn't hold me! I believe I must come back!'

He rose, and I rose too. We grasped each other by the hand again, and as we went out into the rigorous night, everything seemed to be

[1] 'Note from Pocket book'; below, at 'he refolded both', 'Kiss letter' (Wright).

hushed in reverence for him, when he resumed his solitary journey through the snow.

CHAPTER V

ALL this time I had gone on loving Dora harder than ever. * If I may so express it, I was steeped in Dora. I was not merely over head and ears in love with her; I was saturated through and through. I took night walks to Norwood where she lived, and perambulated round and round the house and garden for hours together; looking through crevices in the palings, using violent exertions to get my chin above the rusty nails on the top, blowing kisses at the lights in the windows, and romantically calling on the night to shield my Dora.—*I don't exactly know from what—I suppose from fire—perhaps from mice, to which she had a great objection.* *

Dora had a discreet friend, comparatively stricken in years—*almost of the ripe age of twenty, I should say*—whose name was Miss Mills. Dora called her Julia, and she was the bosom friend of Dora. Happy Miss Mills! *

One day Miss Mills said, 'Dora is coming to stay with me. She is coming the day after to-morrow. If you would like to call, I am sure papa would be happy to see you.' *

I passed three days in a luxury of wretchedness, and at last, arrayed for the purpose at a vast expense, I went to Miss Mills's *fraught with a declaration.*[1]

Mr. Mills was not at home. I didn't expect he would be. Nobody wanted *him.*[2] Miss Mills was at home. Miss Mills would do.

I was shown into a room up-stairs, where Miss Mills and Dora were. Dora's little dog Jip was there. Miss Mills was copying music, and Dora was painting flowers.—*What were my feelings when I recognized flowers I had given her!*

Miss Mills was very glad to see me, and very sorry her papa was not at home: *though I thought we all bore that with fortitude.* Miss Mills was conversational for a few minutes, and then, laying down her pen, got up and left the room.

I began to think I would put it off till to-morrow.[3]

'I hope your poor horse was not tired, when he got home at night from that pic-nic,' said Dora, lifting up her beautiful eyes. 'It was a long way for him.'

[1] Doubly underlined.

[2] *him* italic in *Berg* and in the novel.

[3] This, and the two subsequent 'I began . . .' paragraphs, and 'I saw now . . .', all doubly underlined. They were spoken 'Aside' (Wright).

I began to think I would do it to-day.

'It was a long way for *him*, for *he*[1] had nothing to uphold him on the journey.'

'Wasn't he fed, poor thing?' asked Dora.

I began to think I would put it off till to-morrow.

'Ye-yes, he was well taken care of. I mean he had not the unutterable happiness that I had in being so near you.'

I saw now that I was in for it, and it must be done on the spot.

'I don't know why you should care for being near me,' said Dora, 'or why you should call it a happiness. But of course you don't mean what you say. Jip, you naughty boy, come here!'

I don't know how I did it, but I did it in a moment.[2] I intercepted Jip. I had Dora in my arms. I was full of eloquence. I never stopped for a word. I told her how I loved her. I told her I should die without her. I told her that I idolized and worshipped her.—*Jip barked madly all the time.*

My eloquence increased and I said, if she would like me to die for her, she had but to say the word, and I was ready. I had loved her to distraction every minute, day and night, since I first set eyes upon her. I loved her at that minute to distraction. I should always love her, every minute, to distraction. Lovers had loved before, and lovers would love again; but no lover had ever loved, might, could, would, or should, ever love, as I loved Dora.—*The more I raved, the more Jip barked. Each of us, in his own way, got more mad every moment.*

Well, well! Dora and I were sitting on the sofa by-and-by, quiet enough, and Jip was lying in her lap, winking peacefully at me. *It was off my mind. I was in a state of perfect rapture. Dora and I were engaged.* *[3]

Being poor, I felt it necessary the next time I went to my darling, to expatiate on that unfortunate drawback. I soon carried desolation into the bosom of our joys—not that I meant to do it, but that I was so full of the subject—by asking Dora, without the smallest preparation, *if she could love a beggar?*

'How can you ask me anything so foolish?[4] Love a beggar!'

'Dora, my own dearest, *I*[5] am a beggar!'

'How can you be such a silly thing,' replied Dora, slapping my hand, 'as to sit there, telling such stories? I'll make Jip bite you if you are so ridiculous.'

But I looked so serious, that Dora began to cry. She did nothing but exclaim Oh dear! Oh dear! And oh, she was so frightened! And where

[1] *him . . . he* italic in *Berg* and in the novel.
[2] 'Quickly' (Wright).
[3] These three sentences doubly underlined. The narrative now jumps to ch. 37.
[4] Marginal stage-direction *Sprightly laugh*; contrast the novel's 'pouted Dora' here.
[5] *I* italic in *Berg* and in the novel.

was Julia Mills! And oh, take her to Julia Mills, and go away, please! until I was almost beside myself. *

I thought I had killed her. I sprinkled water on her face. I went down on my knees. I plucked at my hair. I implored her forgiveness. I besought her to look up. *I ravaged Miss Mills's work-box for a smelling-bottle, and in my agony of mind applied an ivory needle-case instead, and dropped all the needles over Dora.*

At last, I got Dora to look at me, with a horrified expression, which I gradually soothed until it was only loving, and her soft, pretty cheek was lying against mine. *

'Is your heart mine still, dear Dora?'

'Oh, yes! Oh, yes, it's all yours. Oh, don't be dreadful!'

I[1] dreadful! To Dora!

'Don't talk about being poor, and working hard! Oh, don't, don't!'

'My dearest love, the crust well earned—'

'Oh, yes; but I don't want to hear any more about crusts. And after we are married, Jip must have a mutton-chop every day at twelve, or he'll die!'

I was charmed with her childish, winning way, and I fondly explained to her that Jip should have his mutton-chop with his accustomed regularity. * †

When we had been engaged some half a year or so, Dora delighted me by asking me to give her that cookery-book I had once spoken of, and to show her how to keep housekeeping accounts, as I had once promised I would. I brought the volume with me on my next visit (*I got it prettily bound, first, to make it look less dry and more inviting*); and showed her an old housekeeping-book of my aunt's, and gave her a set of tablets, and a pretty little pencil-case, and a box of leads, to practise house-keeping with.

But the cookery-book made Dora's head ache, and the figures made her cry. *They wouldn't add up, she said.* So she rubbed them out, and drew little nosegays, and likenesses of me and Jip, all over the tablets. †

Time went on, and at last, here in this hand of mine I held the wedding licence. There were the two names in the sweet old visionary connexion, David Copperfield and Dora Spenlow; and there in the corner was that parental Institution the Stamp-office, looking down upon our union; and there, in the printed form of words, was the Archbishop of Canterbury invoking a blessing on us, *and doing it as cheap as could possibly be expected!* *

I doubt whether two young birds could have known less about keeping house, than I and my pretty Dora did. We had a servant, of course. *She*[2] kept house for us. We had an awful time of it with Mary Anne.

[1] *I* italic in *Berg* and in the novel. [2] *she* doubly underlined.

Her name was Paragon. Her nature was represented to us, when we engaged her, as being feebly expressed in her name. She had a written character, as large as a Proclamation; and, according to this document, could do everything of a domestic nature that I ever heard of, and a great many things that I never did hear of. She was a woman in the prime of life: of a severe countenance; and subject (particularly in the arms) to a sort of perpetual measles. She had a cousin in the Life Guards, with such long legs that he looked like the afternoon shadow of somebody else. She was warranted sober and honest. And I am therefore willing to believe that she was in a fit when we found her under the boiler; and that the deficient teaspoons were attributable to the dustman. She was the cause of our first little quarrel.

'My dearest life.' I said one day to Dora, 'do you think Mary Anne has any idea of time?'

'Why, Doady?'

'My love, because it's five, and we were to have dined at four.'

My little wife came and sat upon my knee, to coax me to be quiet, and drew a line with her pencil down the middle of my nose: *but I couldn't dine off that, though it was very agreeable.*

'Don't you think, my dear, it would be better for you to remonstrate with Mary Anne?'

'Oh no, please! I couldn't, Doady!'

'Why not, my love?'

'Oh, because I am such a little goose, and she knows I am!'

I thought this sentiment so incompatible with the establishment of any system of check on Mary Anne, that I frowned a little. †

'My precious wife, we must be serious sometimes. Come! Sit down on this chair, close beside me! Give me the pencil! There! Now let us talk sensibly. You know, dear;' what a little hand it was to hold, and what a tiny wedding-ring it was to see! 'You know, my love, it is not exactly comfortable to have to go out without one's dinner. Now, is it?'

'N—n—no!' replied Dora, faintly.

'My love, how you tremble!'

'Because I KNOW[1] you're going to scold me.'

'My sweet, I am only going to reason.'

'Oh, but reasoning is worse than scolding! I didn't marry to be reasoned with. If you meant to reason with such a poor little thing as I am, you ought to have told me so, you cruel boy!' †

'Now, my own Dora, you are childish, and are talking nonsense. You must remember, I am sure, that I was obliged to go out yesterday when dinner was half over; and that, the day before, I was made quite unwell

[1] Small capitals in *Berg* and in the novel. Against the passage which follows, Wright has 'Innocently'.

by being obliged to eat underdone veal in a hurry; to-day, I don't dine at all—and I am afraid to say how long we waited for breakfast—and *then*[1] the water didn't boil. I don't mean to reproach you, my dear, but this is not comfortable.'

'I wonder, I do, at your making such ungrateful speeches. When you know that the other day, when you said you would like a little bit of fish, I went out myself, miles and miles, and ordered it, to surprise you.'

'And it was very kind of you, my own darling, and I felt it so much that I wouldn't on any account have mentioned that you bought a salmon—which was too much for two. Or that it cost one pound six—which was more than we can afford.'

'You enjoyed it very much,' sobbed Dora. 'And you said I was a mouse.'

'And I'll say so again, my love, a thousand times!' *

I said it a thousand times, and more, and went on saying it until Mary Anne's cousin deserted into our coal-hole, and was brought out, to our great amazement, by a piquet of his companions in arms, who took him away handcuffed, in a procession that covered our front-garden with disgrace. †

Everybody we had anything to do with, seemed to cheat us. Our appearance in a shop was a signal for the damaged goods to be brought out immediately. If we bought a lobster, it was full of water. All our meat turned out tough, and there was hardly any crust to our loaves. * As to the *washerwoman*[2] pawning the clothes, and coming in a state of penitent intoxication to apologize, I suppose that might have happened several times to anybody. Also the chimney on fire, the parish engine, and perjury on the part of the *Beadle*. But I apprehend we were personally unfortunate in our *page*: whose principal function was to quarrel with the cook and who lived in a hail of saucepan-lids.[3] We wanted to get rid of him, but he was very much attached to us, and wouldn't go, * until one day he stole Dora's watch, and spent the produce (he was always a weak-minded boy) in riding up and down between London and Uxbridge outside the coach. He was taken to the Police Office, on the completion of his fifteenth journey; when four-and-sixpence, and a second-hand fife which he couldn't play, were found upon his person. *

He was tried and ordered to be transported. Even then he couldn't be quiet, but was always writing us letters; and he wanted so much to see Dora before he went away, that Dora went to visit him, and fainted when she found herself inside the iron bars. I had no peace of my life until he was expatriated, and made (as I afterwards heard) a shepherd of, 'up the country' somewhere; I have no geographical idea where.

[1] *then* italic in *Berg* and in the novel.

[2] *Washerwoman*, and *Beadle* and *page* (below), all doubly underlined.

[3] In *Berg*, Dickens deleted 'and who' and may have intended to delete the rest of the sentence (as is done in *T & F*). The 'page' episode is from ch. 48.

'I am very sorry for all this, Doady,' said Dora. * 'Will you call me a name I want you to call me?'

'What is it, my dear?'

'It's a stupid name—Child-wife. When you are going to be angry with me, say to yourself "it's only my Child-wife."[1] When I am very disappointing, say, "I knew, a long time ago, that she would make but a Child-wife." When you miss what you would like me to be, and what I should like to be, and what I think I never can be, say, "Still my foolish Child-wife loves me." For indeed I do.' *

I invoke the innocent figure that I dearly loved, to come out of the mists and shadows of the Past, and to turn its gentle head towards me once again, and to bear witness that it was made happy by what I answered.

CHAPTER VI

I HEARD a footstep on the stairs one day. I knew it to be Mr. Peggotty's. It came nearer, nearer, rushed into the room.

'Mas'r Davy, I've found her! I thank my Heavenly Father for having guided of me in His own ways to my darling!'[2] †

'You have made up your mind as to the future, good friend?'

'Yes, Mas'r Davy, theer's mighty countries, fur from heer. Our future life lays over the sea.' *

As he gave me both his hands, hurrying to return to the one charge of his noble existence, I thought of Ham and who would break the intelligence to him? Mr. Peggotty thought of everything. He had already written to the poor fellow, and had the letter in the pocket of his rough coat, ready for the post. I asked him for it, and said I would go down to Yarmouth, and talk to Ham myself before I gave it him, and prepare him for its contents. He thanked me very earnestly, and we parted, with the understanding that I would go down by the Mail that same night. In the evening I started.

'Don't you think that,' I asked the coachman,[3] in the first stage out

[1] 'Feelingly' (Wright). This final dialogue comes from ch. 44.

[2] Dickens made successive attempts to shorten Peggotty's long narrative (from ch. 51), which followed here. There are many deletions in *Berg*; then pp. 81–4 were stuck together with stamp-edging; finally Dickens jettisoned the whole narrative and, on p. 80, indicated a jump to p. 88, and again used stamp-edging to eliminate the unwanted pages.

[3] 'Looking up at ceiling' (Wright). The storm-scene which follows was greatly admired (see headnote). Accounts about its rendering differ—though performances doubtless differed, too. Thus, one critic wrote: it 'was a brilliant bit of elocution; intensely powerful, yet perfectly quiet; subdued, though telling, in every phrase' (*Scotsman*, 28 November 1861). Another did not find it quiet or subdued: 'it was, if anything, a little too dramatic for the platform, and . . . scarcely so easy and natural as many of the other passages' (*Carlisle Journal*, 13 December 1861). Wright puts 'Exaggerated' at several points in the final pages.—The *Berg* text is extremely heavily cut here. The narrative jumps from ch. 51 to ch. 55.

of London, 'a very remarkable sky? I don't remember to have ever seen one like it.'

'Nor I. That's wind, sir. There'll be mischief done at sea before long.'

It was a murky confusion of flying clouds tossed up into most remarkable heaps, through which the wild moon seemed to plunge headlong, as if, in a dread disturbance of the laws of nature, *she had lost her way*. There had been a wind all day; and it was rising then, with an extraordinary great sound. In another hour it had much increased, and the sky was more overcast, and it blew hard.

But, as the night advanced, it came on to blow, harder and harder.[1] † I had been in Yarmouth when the seamen said it blew great guns, but I had never known the like of this, or anything approaching to it. * †

The tremendous sea itself, when I came to my journey's end, confounded me. As the high watery walls came rolling in, and tumbled into surf, † I seemed to see a rending and upheaving of all nature.

Not finding Ham among the people whom this memorable wind had brought together on the beach, I made my way to his house. I learned that he had gone, on a job of shipwright's work, some miles away, but that he would be back to-morrow morning, in good time.

So, I went back to the inn; and when I had washed and dressed, and tried to sleep, but in vain, it was late in the afternoon. I had not sat five minutes by the coffee-room fire, when the waiter coming to stir it, told me that two colliers had gone down, with all hands, a few miles off; and that some other ships had been seen labouring hard in the Roads, and trying, in great distress, to keep off shore. Mercy on them, and on all poor sailors, said he, if we had another night like the last! *

I could not eat, I could not sit still, I could not continue stedfast to anything. My dinner went away almost untasted, and I tried to refresh myself with a glass or two of wine. In vain.[2] I walked to and fro, tried to read an old gazetteer, listened to the awful noises: looked at faces, scenes, and figures in the fire. At length the ticking of the *undisturbed* clock on the wall, tormented me to that degree that I resolved to go to bed.

For hours, I lay in bed listening to the wind and water; imagining, now, that I heard shrieks out at sea; now, that I distinctly heard the firing of signal guns; now, the fall of houses in the town. At length, my restlessness attained to such a pitch, that I hurried on my clothes, and went down-stairs. In the large kitchen, all the inn servants and some other watchers were clustered together. One man asked me when I went

[1] Dickens deleted too much here, and it is uncertain whether he meant to *stet* 'it came on to blow, harder and harder' or a similar phrase, later, 'the wind blew harder and harder'.

[2] 'Dramatic. Impatiently' (Wright).

in among them whether I thought the souls of the collier-crews who had gone down, were out in the storm?

There was a dark gloom in my lonely chamber, when I at length returned to it; but I was tired now, and, getting into bed again, fell into the depths of sleep until broad day; when I was aroused, at eight or nine o'clock, by some one knocking and calling at my door.

'What is the matter?'

'A wreck! Close by!'[1]

'What wreck?'

'A schooner, from Spain or Portugal, laden with fruit and wine. Make haste, sir, if you want to see her! It's thought, down on the beach, she'll go to pieces every moment.'

[2]I wrapped myself in my clothes as quickly as I could, and ran into the street, where numbers of people were before me, all running in one direction—to the beach.

When I got there,—in the difficulty of hearing anything but wind and waves, and in the crowd, and the unspeakable confusion, and my first breathless efforts to stand against the weather, I was so confused that I looked out to sea for the wreck, and saw nothing but the foaming heads of the great waves. A boatman laid a hand upon my arm, and pointed. Then, I saw it, close in upon us!

One mast was broken short off, six or eight feet from the deck, and lay over the side, entangled in a maze of sail and rigging; and all that ruin, as the ship rolled and beat—which she did with a violence quite inconceivable—beat the side as if it would stave it in. Some efforts were being made, to cut this portion of the wreck away; for, as the ship, which was broadside on, turned towards us in her rolling, I plainly descried her people at work with axes—especially one active figure with long curling hair.[3] But, a great cry, audible even above the wind and water, rose from the shore; *the sea, sweeping over the wreck, made a clean breach, and carried men, spars, casks, planks, bulwarks, heaps of such toys, into the boiling surge.*

The second mast was yet standing, with the rags of a sail, and a wild confusion of broken cordage flapping to and fro. The ship had struck once, the same boatman said, *and then lifted in and struck again.* I understood him to add that she was parting amidships. As he spoke, there was another great cry of pity from the beach. *Four men arose with the wreck out of the deep, clinging to the rigging of the remaining mast; uppermost, the active figure with the curling hair.*[4]

[1] 'High voice' (Wright).

[2] Marginal stage-direction *Quick.*

[3] The final phrase—*especially one active figure with long curling hair*—doubly underlined. Later references to Steerforth are similarly distinguished. Wright has, against the rest of this paragraph, 'Quick'.

[4] *Four men arose* and *uppermost, the active figure with the curling hair* doubly underlined.

There was a bell on board; and as the ship rolled and dashed, this bell rang; and its sound, the knell of those unhappy men, was borne towards us on the wind. Again we lost her, and again she rose. Two of the four men were gone.[1] †

I noticed that some new sensation moved the people on the beach, and I saw them part, and Ham come breaking through them to the front.

Instantly, I ran to him, for I divined that he meant to wade off with a rope. I held him back with both arms; and implored the men not to listen to him, not to let him stir from that sand!

Another cry arose; and we saw the cruel sail, with blow on blow, beat off the lower of the two men, and fly up in triumph round the active figure left alone upon the mast.[2]

Against such a sight, and against such determination as that of the calmly desperate man who was already accustomed to lead half the people present, I might as hopefully have entreated the wind.

I was swept away to some distance, where the people around me made me stay; urging, as I confusedly perceived, that he was bent on going, with help or without, and that I should endanger the precautions for his safety by troubling those with whom they rested. I saw hurry on the beach, and men running with ropes, and penetrating into a circle of figures that hid him from me. Then, I saw him standing alone, in a seaman's frock and trowsers: a rope in his hand: another round his body: and several of the best men holding to the latter.

The wreck was breaking up. I saw that she was parting in the middle, and that the life of the solitary man upon the mast hung by a thread.[3] *He had a singular red cap on, not like a sailor's cap, but of a finer colour; and as the few planks between him and destruction rolled and bulged, and as his death-knell rung, he was seen by all of us to wave this cap.* I saw him do it now, *and thought I was going distracted, when his action brought an old remembrance to my mind of a once dear friend—the once dear friend—Steerforth.*[4]

Ham watched the sea, until there was a great retiring wave; when he dashed in after it, and in a moment was buffeting with the water, rising with the hills, falling with the valleys, lost beneath the foam: † *borne in towards the shore, borne on towards the ship. At length he neared the wreck. He was so near, that with one more of his vigorous strokes he would be clinging to it,—when, a high green vast hill-side of water, moving on*

[1] *Two men were gone* doubly underlined; *of the four* is a marginal insertion.

[2] *beat off the lower of the two men* and *round the active figure left alone upon the mast* doubly underlined.

[3] *and that the life . . . by a thread* doubly underlined.

[4] *and thought I was* to the end of the paragraph doubly underlined; *the* italic in *Berg* and in the novel. Wright notes 'Quick' against '*and thought I was . . .*', and a 'Pause' before '*Steerforth*'.

shoreward, from beyond the ship, he seemed to leap up into it with a mighty bound—and the ship was gone![1]

They drew him to my very feet—insensible—dead.[2] He was carried to the nearest house; and every means of restoration were tried; but he had been beaten to death by the great wave, and his generous heart was stilled for ever.

As I sat beside the bed,[3] when hope was abandoned and all was done, a fisherman, who had known me when Emily and I were children, and ever since, whispered my name at the door.

'Sir, will you come over yonder?'

The old remembrance that had been recalled to me, was in his look, and I asked him:

'Has a body come ashore?'

'Yes.'[4]

'Do I know it?'

He answered nothing. But, he led me to the shore. And on that part of it where she and I had looked for shells, two children—on that part of it where some lighter fragments of the old boat, blown down last night, had been scattered by the wind—among the ruins of the home he had wronged—I saw him[5] lying with his head upon his arm, as I had often seen him lie at school.

THE END OF THE READING

[1] 'Long pause' after *a mighty bound* (Wright); *and the ship was gone!* doubly underlined.

[2] Marginal stage-direction *Low*. Wright's note is 'Solemn and slow'.

[3] Marginal stage-direction *Lower*.

[4] Here 'Mr. Dickens displayed his dramatic power in a very remarkable manner. The tone in which David, knowing what the answer will be, and yet dreading to hear it, asks, "Has a body come ashore?"—strikes to the heart of every person within reach of his voice. And the answer! In the book it is simply "yes"; but Mr. Dickens, in the person of the old fisherman, does not speak,—he only bows his head; and in that simple action conveys the whole story which the lips cannot speak. Acting more impressive than this we have never witnessed. The whole audience felt its power, and the hush that fell upon the room was for the moment almost painful' (*New York Times*, 11 December 1867).

[5] Wright deletes 'him' and substitutes 'Steerforth'. 'Do you cry when you read aloud?' a twelve-year-old American girl asked Dickens, when she waylaid him on a train. 'We all do in our family. And we never read about Tiny Tim, or about Steerforth when his body is washed up on the beach, on Saturday nights, or our eyes are too swollen to go to Sunday School.'—'Yes, I cry when I read about Steerforth,' Dickens answered quietly (Kate Douglas Wiggin, *A Child's Journey with Dickens* (Boston and New York, 1912), p. 27).

NICHOLAS NICKLEBY AT THE YORKSHIRE SCHOOL

HAVING completed the *Copperfield* script, Dickens turned to *Nickleby*, and these two new items opened his provincial tour in October 1861. 'I think Nickleby tops all the readings', Dickens reported after its première. 'Somehow it seems to have got in it, by accident, exactly the qualities best suited to the purpose, and it went last night not only with roars, but with a general hilarity and pleasure that I have never seen surpassed' (*N*, iii. 246). The audience's merriment during the letter-reading scene could even make Dickens himself collapse with laughter, though he later admitted that it was a long time before he enjoyed giving this Reading—'But I got better, as I found the audience always taking to it' (*N*, iii. 353).

Unlike *Copperfield*, it was never a two-hour reading, but was paired with *The Trial* or another short item. During the 1866 season, however, it was paired with a long item, *Dr. Marigold*, and for this purpose Dickens prepared a 'short-time' version. The full 1861 Version (*Berg*) consists of four Chapters: Chapter I, mainly from ch. 7 of the novel, describes Nicholas's journey with Mr. Squeers from London to Yorkshire. Chapters II (the school) and III (Fanny Squeers's tea-party) correspond to chs. 8 and 9; Chapter IV (Nicholas thrashes Squeers, leaves Dotheboys Hall, and is joined by Smike) derives from chs. 12 and 13, with a paragraph from ch. 64 ('Dotheboys Hall breaks up for ever'). The 'short-time' version (*Suzannet*) omits Chapter III and differs in other minor respects from *Berg*; a facsimile of it was published by the Ilkley Literature Festival, 1973. Dickens later returned to the *Berg* text (reprinted here), though he incorporated into his performances a few successful passages unique to *Suzannet*.

The Reading was usually advertised under the title of *Nicholas Nickleby at Mr. Squeers's School*, the naming of Squeers being a reminder of the dominant character in the piece. Squeers was very droll, particularly in the letter-reading scene, but audiences' righteous indignation against him mounted, and the moment for which they were perceptibly waiting was Nicholas's intervention to protect Smike. His 'battle with the tyrant' was 'narrated with a stirring manliness and vigour that told greatly'; audiences 'actually exulted in the punishment' that Squeers received, and cheered Nicholas as the thrashing proceeded. The best comic moment was Fanny Squeers's tea-party; everyone delighted in Fanny's lisp and John Browdie's blunt heartiness, expressed in a Yorkshire accent that was at least found convincing in Lancashire. The Reading contained many good character-parts, and Dickens was said to have assumed them well, though opinions differed over his rendering of Smike. Most people found this very affecting: applause was general at the 'tender touches', and tears flowed at the conclusion. 'John Browdie captivated the audience most', reported the *Birmingham Gazette* (3 April 1869), 'but we must give the palm to the far more difficult character of poor Smike, which, in any other hands, of very doubtful success, in Mr. Dickens's is a gem of pure pathos'. Others, however, such as Kate Field, found Smike monotonous, whiney and unnatural—'not poorly done, but it can be better done' (pp. 61–2).

W. M. Wright annotated his copy of *Nickleby* very fully, often noting the gestures which Dickens's numerous stage-directions enjoined: his notes here, indeed, attest his accuracy as an observer. His account of the characters' voices is useful: 'Natural voice for Nickleby', 'Mrs. Cluppins's voice' for Mrs. Squeers, 'Quick slight voices' for the boys at their class in 'English spelling and philosophy', 'High dismal voice' and 'Breathlessly' for Smike, and 'High laugh' for Browdie. Squeers pronounces 'put' as in 'putty', and is apt to aspirate words like 'angel': he is evidently a Londoner.

Nicholas Nickleby

4 Chapters[1]

CHAPTER I

NICHOLAS NICKLEBY, in the nineteenth year of his age, arrived at eight o'clock of a November morning at the sign of the Saracen's Head, Snow Hill, London, *to join Mr. Squeers, the cheap—the terribly cheap—Yorkshire schoolmaster.*[2] Inexperienced, sanguine, and thrown upon the World with no adviser, and his bread to win, he had engaged himself to Mr. Squeers as his *scholastic assistant*, on the faith of the following advertisement in the London papers:

'EDUCATION.—At Mr. Wackford Squeers's Academy, Dotheboys Hall, at the delightful village of Dotheboys, near Greta Bridge in Yorkshire, Youth are boarded, clothed, booked, furnished with pocket-money, provided with all necessaries, instructed in all languages living and dead, mathematics, orthography, geometry, astronomy, trigonometry, the use of the globes, algebra, single stick (if required), writing, arithmetic, fortification, and every other branch of classical literature. Terms, twenty guineas per annum. *No extras, no vacations, and diet unparalleled.*[3] Mr. Squeers is in town, and attends daily, from one till four, at the Saracen's Head, Snow Hill. N.B. An able assistant wanted. *Annual Salary £5. A Master of Arts would be preferred.*' *

Mr. Squeers was standing by one of the coffee-room fireplaces, and his appearance was not prepossessing. He had but one eye, and the popular prejudice runs in favour of two. The blank side of his face was much puckered up, which gave him a sinister appearance, especially when he smiled; at which times his expression bordered on the villanous. He wore a white neckerchief with long ends, and a scholastic suit of black; but his coat-sleeves being a great deal too long, and his trousers a great

[1] *IN FOUR* deleted, *3* inserted and then deleted, and *4* inserted. At one stage, Dickens made this (full-length) version of *Nickleby* a three-chapter Reading, by running Chapters I and II into one. Later, he reverted to the four-chapter division. Wright notes Dickens's brief introduction: 'I am to have the pleasure of reading to you first tonight . . .' The opening paragraph is a summary, not in the novel.

[2] Doubly underlined, as is *scholastic assistant* below. The words 'cheap—the terribly cheap—' and the phrase from 'Inexperienced' to 'bread to win' do not appear in *T & F*, so are a late manuscript addition.

[3] *and diet unparalleled* doubly underlined. Dickens 'laid particular stress' upon this sentence (Kent, p. 192). The underlining in the *Suzannet* 'short-time' copy is generally similar to that in *Berg*, as here; only significant differences will be noted. Against this paragraph, Wright notes: 'Holding handkerchief both hands'.

deal too short, *he appeared ill at ease in his clothes, and as if he were in a perpetual state of astonishment at finding himself so respectable.* *

The learned gentleman had before himself, a breakfast of coffee, hot toast, and cold round of beef; but he was at that moment intent on preparing another breakfast for five little boys.

[1]'This is twopenn'orth of milk is it, waiter?' said Mr. Squeers, looking down into a large mug.

'That's twopenn'orth, sir.'

'What a rare article milk is, to be sure, in London! Just fill that mug up with lukewarm water, William, will you?'

'To the wery top, sir? Why, the milk will be drownded.'

'*Serve it right for being so dear.* You ordered that thick bread and butter for three, did you?'

'Coming directly, sir.'

[2]'You needn't hurry yourself; there's plenty of time. *Conquer your passions, boys, and don't be eager after vittles.*' As he uttered this moral precept, Mr. Squeers took a large bite out of the cold beef, and recognized Nicholas.

[3]'Sit down, Mr. Nickleby. Here we are, a breakfasting you see!' *Nicholas did not see that anybody was breakfasting, except Mr. Squeers.*[4]

'Oh! that's the milk and water, is it, William?[5] Here's richness! Think of the many beggars and orphans in the streets that would be glad of this, little boys. When I say number one, the boy on the left hand, nearest the window, may take a drink; and when I say number two, the boy next him will *go in*,[6] and so till we come to number five. Are you ready?'

'Yes, sir.'

[7]'Keep ready till I tell you to begin. Subdue your appetites, and you've conquered human natur. This is the way we inculcate strength of mind, Mr. Nickleby.'

Nicholas murmured something in reply; and the little boys remained in torments of expectation.

[1] Marginal stage-direction *The Mug* (in both margins). 'The mug is *not* on that desk, but it seems to be, as Mr. Squeers looks into it and gives his order' (Field, p. 55).

[2] Marginal stage-directions *Chuckle* (in the left margin) and *Chuckling* (in the right margin).

[3] Marginal stage-direction *Breakfasting*. Just below, opposite the paragraph (deleted in *Berg*) beginning 'At this fresh mention . . .', there is the stage direction *Stir*. This piece of business was retained: 'Stirring all time' (Wright; Kent, p. 143).

[4] *not* doubly underlined in *Berg*; whole sentence doubly underlined in *Suzannet*. Wright, however, deletes the sentence.

[5] Marginal stage-direction *Smack*.

[6] *go in* doubly underlined. The boys' reply, 'Yes, sir', came in a 'High voice' (Wright).

[7] Marginal stage-directions *Chuckle* (in the left margin) and *Chuckling* (in the right margin).

[1]'Thank God for a good breakfast. Number one may take a drink.'

Number one seized the mug ravenously, and had just drunk enough to make him wish for more, when Mr. Squeers gave the signal for number two, who gave up at the like interesting moment to number three; and the process was repeated until the milk and water terminated with number five.

'And now,' said the schoolmaster, dividing the bread and butter for three into five portions, 'you had better look sharp with your breakfast, for the coach-horn will blow in a minute or two, and then every boy leaves off.'

The boys began to eat voraciously, while the schoolmaster (who was in high good humour after his meal) picked his teeth with a fork, and looked on. In a very short time, the horn was heard.

[2]'I thought it wouldn't be long' said Squeers, jumping up and producing a little basket; 'put what you haven't had time to eat, in here, boys! You'll want it on the road!' *

They certainly *did*[3] want it on the road, and very much, too; for the journey was long, the weather was intensely cold, a great deal of snow fell from time to time, and the wind was intolerably keen. Mr. Squeers got down at almost every stage—*to stretch his legs, he said*[4]—and as he always came back with a very red nose, and composed himself to sleep directly, *the stretching seemed to answer*. * It was a long journey, but the longest lane has a turning at last, and late in the night the coach put them down at a lonely roadside Inn where they found in waiting two labouring men, a rusty pony-chaise, and a cart.

'Put the boys and the boxes into the cart, and this young man and me will go on in the chaise. Get in, Nickleby.'

Nicholas obeyed. Mr Squeers with some difficulty inducing the pony to obey too, they started off, leaving the cart-load of infant misery to follow at leisure.

[5]'Are you cold, Nickleby?'

'Rather, sir, I must say.'

'Well, I don't find fault with that. It's a long journey this weather.'

'Is it much further to Dotheboys Hall, sir?'

'About three mile. But you needn't call it a Hall down here.'

Nicholas coughed, as if he would like to know why.

[1] Marginal stage-direction *Wiping mouth.* Wright, who notes this piece of business, writes at the bottom of this page of *T & F*, 'Slow, monotonous', and at the top of the next 'Reading in a low key'.

[2] Marginal stage-direction *Glee.* At the end of this paragraph there is a long deletion of the novel's text.

[3] *did* italic in *Berg*.

[4] Doubly underlined.

[5] Marginal stage-direction *Driving*. Against the dialogue following, Wright notes: 'Natural voice for Nickleby'.

'The fact is, it ain't a Hall.'

'Indeed!'

'No. We call it a Hall up in London, because it sounds better, but they don't know it by that name in these parts. A man may call his house an island if he likes; there's no act of Parliament against that, I believe?'

Squeers eyed him at the conclusion of this little dialogue, and finding that he had grown thoughtful, contented himself with lashing the pony until they reached their journey's end,[1] * when he ushered him into a small parlour scantily furnished, where they had not been a couple of minutes, when a female bounced into the room, and, seizing Mr. Squeers by the throat, gave him two loud kisses: *one close after the other, like a postman's knock*. This lady was of a large raw-boned figure, about a head taller than Mr. Squeers, and was dressed in a dimity night-jacket; with her hair in papers and a dirty nightcap. (She was accustomed to boast that she was no grammarian, thank God; and also that she had tamed a high spirit or two in her day. Truly, in conjunction with her worthy husband, she had broken many and many a one.)[2]

'How is my Squeery?'

'Quite well, my love. How's the cows?'

'The cows is all right, every one of 'em.'

'And the pigs?'

'The pigs is as well as they was when you went away.'

'Come! That's a blessing! The boys are all as they were, I suppose?'

'Oh, yes, the boys is well enough. Only that young Pitcher's had a fever.'

'No! Damn that chap, he's always at something of that sort.'

Pending these endearments, Nicholas had stood, awkwardly enough, in the middle of the room: not very well knowing whether he was expected to retire into the passage. He was now relieved from his perplexity by Mr. Squeers.

'This is the new young man, my dear.'[3]

Here, a young servant-girl brought in some cold beef, and this being set upon the table, *a boy addressed by the name of Smike appeared with a jug of ale*.[4]

[1] *T & F* contains a short passage here (deleted in *Berg*) describing the arrival at Dotheboys Hall. Smike had a brief first entrance here, and there is in *Berg* a marginal stage-direction *Smike*. Nicholas's lengthy first impressions of Smike here had been omitted, however, before *Berg* was printed.

[2] The bracketed sentences, drawn from ch. 9, are added in the margin of *Berg*.

[3] Dickens's stage-business here may be surmised from Mrs. Squeers's 'Oh!' which was printed, but deleted, in *Berg*. Dickens had there omitted the following words, 'replied Mrs. Squeers, nodding her head at Nicholas, and eyeing him coldly from top to toe': he performed, instead of reading, such sentences.

[4] Marginal stage-direction *Smike*. The phrase referring to him is doubly underlined. Against the following paragraph, Wright notes: 'As if pulling letters out of pocket'.

Mr. Squeers was emptying his greatcoat pockets of letters and other small documents he had brought down. The boy glanced, with an anxious and timid expression, at the papers, as if with a sickly hope that one among them might relate to him. The look was a very painful one, and went to Nicholas's heart at once; for it told a long and very sad history.

It induced him to consider the boy more attentively, and he was surprised to observe the extraordinary mixture of garments which formed his dress. Although he could not have been less than eighteen or nineteen, and was tall for that age, he wore a skeleton suit, such as was then[1] usually put upon a very little boy. In order that the lower part of his legs might be in keeping with this singular dress, he had a very large pair of boots, originally made for tops, which might have been once worn by some stout farmer, but were now too patched and tattered for a beggar. God knows how long he had been there, but he still wore a tattered child's frill, only half concealed by a coarse, man's neckerchief. He was lame; and as he feigned to be busy in arranging the table, he glanced at the letters with a look so keen, and yet so dispirited and hopeless, that Nicholas could hardly bear to watch him.

'What are you bothering about there, Smike?' cried Mrs. Squeers; 'let the things alone, can't you?'

'Eh! Oh! it's you, is it?'

'Yes, sir. Is there—'

'Well! What are you stammering at?'

'Have you—did anybody—has nothing been heard—about me?'[2]

'Devil a bit, not a word, and never will be. Now this is a pretty sort of thing, isn't it, that you should have been left here, all these years, and no money paid after the first six—nor no notice taken, nor no clue to be got who you belong to? It's a pretty sort of thing that I should have to feed a great fellow like you, and never hope to get a penny for it, isn't it?'

The boy put his hand to his head as if he were making an effort to recollect something, and then, looking vacantly at his questioner, gradually broke into a smile, and limped away.

'I'll tell you what, Squeers, I think that young chap's turning silly.'

[3]'I hope not, for he's a handy fellow out of doors, and worth his meat and drink, any way. Hows'ever, I should think he'd have wit enough for us, if he *was*[4] silly. But come! Let's have supper, for I'm hungry and tired, and want to get to bed.'[5]

[1] 'was then' substituted in *Berg* for 'is'—an example of the up-dating sometimes needed in the Readings.

[2] 'Breathlessly' (Wright). Against the next paragraph (Squeers's speech), marginal stage direction *Bullying*.

[3] Marginal stage-direction *Yawning*. 'Squeers's right eye closed' (Wright).

[4] *was* italic in *Berg*.

[5] The printed text of Chapter I of *Berg* (and *Suzannet*) ends here. The remainder is written in, or pasted in from a cut-up copy of the novel.

This reminder brought in an exclusive steak for Mr. Squeers, * and Nicholas had a tough bit of cold beef. Mr. Squeers then took a Bumper of hot brandy and water of a stiff nature, and Mrs. Squeers made the new young man the Ghost of a small glassful of that compound.

Then: Mr. Squeers yawned again, and opined that it was time to go to bed;[1] upon which signal, Mrs. Squeers and the girl dragged in a straw mattress and a couple of blankets, and arranged them into a couch for Nicholas.

'We'll put you into your regular bed-room to-morrow, Nickleby. Let me see! Who sleeps in Brooks's bed, my dear?'

'In Brooks's? There's Jennings, little Bolder, Graymarsh, and what's his name.'

'So there is. Yes! Brooks is full. There's a place somewhere, I know, but I can't at this moment call to mind where. However, we'll have that all settled to-morrow. Good night, Nickleby. Seven o'clock in the morning, mind.'

'I shall be ready, sir. Good night.'[2]

'I don't know, by the bye, whose towel to put you on; but if you'll make shift with something to-morrow morning, Mrs. Squeers will arrange that, in the course of the day. My dear, don't forget.'

Mr. Squeers then nudged Mrs. Squeers[3] to bring away the brandy bottle, lest Nicholas should help himself in the night; and the lady having seized it with great precipitation, they retired together. *

CHAPTER II[4]

A ride of two hundred and odd miles in winter weather, is a good softener of a hard bed. Perhaps it is even a sweetener of dreams, for those which came to Nicholas, and whispered their airy nothings, were of a happy kind. He was making his fortune very fast indeed—in his sleep—when the faint glimmer of a candle shone before his eyes, and Mr. Squeers's voice admonished him that it was time to get up.

[5]'Past seven, Nickleby!'

'Has morning come already?'

'Ah! that has it, and ready iced too. Now, Nickleby, come; tumble up!'

[1] 'Rubbing forehead'; at the end of this paragraph, 'Action with both hands' (Wright).

[2] An unwritten exchange followed here: Squeers enquired, 'Do you wash?' to which Nicholas replied, 'Occasionally' (Field, p. 56). Another elaboration of the ablutions at Dotheboys Hall appears in *Suzannet*: 'make shift with something', in Squeers's next speech, becomes 'make shift with your pocket handkerchief.'

[3] 'With elbow' (Wright).

[4] 'No pause between chapters' (Wright).

[5] Marginal stage-direction *Waking cold*. The first paragraph is cut in *Suzannet*.

Nicholas 'tumbled up,' and proceeded to dress himself by the light of Mr. Squeers's candle.

'Here's a pretty go,' said that gentleman; 'the pump's froze.'

'Indeed!'

'Yes. You can't wash yourself this morning.'

'Not wash myself!'

'Not a bit of it.[1] So you must be content with giving yourself a dry polish till we break the ice in the well, and get a bucketful out for the boys. Don't stand staring at me, but look sharp!'

Nicholas huddled on his clothes; * and Squeers, arming himself with his cane, led the way across a yard, to a door in the rear of the house.

'There! This is our shop, Nickleby!'[2]

A bare and dirty room, with a couple of windows, of which a tenth part might be of glass, the remainder being stopped up with old copybooks and paper. A couple of old desks, cut and notched, and inked, and damaged, in every possible way; two or three forms; a detached desk for Squeers; another for his assistant. Walls so discoloured, that it was impossible to tell whether they had ever been touched with paint or whitewash.[3] *†

The boys took their places and their books, *of which latter commodity the average might be about one to a dozen*[4] *learners*. A few minutes having elapsed, *during which Mr. Squeers looked very profound, as if he had a perfect apprehension of what was inside all the books, and could say every word of their contents by heart if he chose*, that gentleman called up the first class.

There ranged themselves in front of the schoolmaster's desk, a dozen haggard scarecrows, out at knees and elbows, one of whom placed a filthy book beneath his learned eye.

'This is the first class in English spelling and philosophy, Nickleby. We'll get up a Latin one, and hand that over to you. Now, then, where's the first boy?'

[1] 'Rubbing hands' (Wright).

[2] 'Open hand to right with a flourish' (Wright).

[3] *T & F* prints here the paragraph, beginning 'But the pupils!' which is deleted—obviously later—in *Berg*. (Wright, too, deletes it.) It is left standing in *Suzannet*. This deletion is interesting, for it contains the grimmest account of the boys' plight, and had been very effective when read: e.g., the *Western Daily Mercury*, 7 January 1862, noted that 'the deeply pathetic passages' in this Reading 'fell with their full force . . . effectually upon the ear, through the slow and simple cadences of the reader. Thus nothing could be finer than those passages descriptive of the wretched misery of the poor boys at Dotheboys Hall as they were first seen by Nickleby . . .' The brimstone and treacle, and the porridge, episodes which followed this description were not printed in *Berg* (or *Suzannet*); one reviewer recalled Dickens's including them in a performance a few years back—but was probably misremembering (Edinburgh *Courant*, 15 December 1868). As was noted above, in the headnote, the episode does appear in *Gimbel*.

[4] *a dozen* substituted for *eight* in manuscript: the usual rounding up and amplifying of figures in the Readings texts. Similarly, in the next paragraph, 'half a dozen' becomes 'a dozen'.

'Please, sir, he's cleaning the back parlour window,' said the temporary head of the philosophical class.

'So he is, to be sure. We go upon the practical mode of teaching, Nickleby; the regular education system. C-l-e-a-n, clean, verb active, to make bright, to scour. W-i-n, win, d-e-r, der, winder, a casement.[1] *When the boy knows this* out of book, *he goes and does it.*[2] It's just the same principle as the use of the globes. Where's the second boy?'

'Please, sir, he's weeding the garden.'

'To be sure. So he is. B-o-t, bot, t-i-n, tin, bottin, n-e-y, bottinney, noun substantive, a knowledge of plants. When he has learned that bottiney means a knowledge of plants, *he goes and knows 'em.* That's our system, Nickleby. Third boy, what's a horse.'

'A beast, sir.'

'So it is. Ain't it, Nickleby?'

'I believe there is no doubt of that, sir.'[3]

'Of course there ain't. A horse is a quadruped, and quadruped's Latin for beast. As you're perfect in that, boy, go and look after *my*[4] horse, and rub him down well, or I'll rub you down. The rest of the class go and draw water up, till somebody tells you to leave off, for it's washing-day to-morrow, and they want the coppers filled.'

So saying, he dismissed the first class *to their experiments in practical philosophy*, and eyed Nicholas with a look, half cunning and half doubtful, *as if he were not altogether certain what he might think of him by this time.*

[5]'That's the way we do it, Nickleby, and a very good way it is. Now, just take them fourteen little boys and hear them some reading, because, you know, you must begin to be useful, and idling about here, won't do.' *

It was Mr. Squeers's custom to call the boys together, and make a sort of report, after every half-yearly visit to the metropolis. So, in the afternoon, the boys were recalled from house-window, garden, stable, and cow-yard, and the school were assembled in full conclave.

[6]'Let any boy speak a word without leave,' said Mr. Squeers, 'and I'll take the skin off that boy's back.'

Death-like silence immediately prevailed.

[1] Dickens increased the absurdity of Mr. Squeers's pedagogy, though these improvements do not appear in *Berg* or *Suzannet*. Kate Field (p. 57) records the happy inventions—'. . . winder, *preposition*, a casement . . . bottiney, *adjective*, a knowledge of plants . . .' and, best of all, 'and quadruped's Latin, *or Greek, or Hebrew, or some other language that's dead and deserves to be*, for beast . . .'

[2] *he goes and does it* doubly underlined, as is *he goes and knows 'em* below.

[3] 'With a sigh' (Wright).

[4] *my* italic in the novel and in *Berg*. Wright notes against this speech, 'Threatening with forefinger': also that Dickens altered 'coppers' to 'boilers'.

[5] Wright deletes this paragraph.

[6] Marginal stage-direction *Slapping desk*. Kent (p. 145) describes it as a 'ferocious slash on the desk with his cane'.

'Boys, I've been to London, and have returned to my family and you, as strong and as well as ever.'

The boys gave three feeble cheers at this refreshing intelligence. Such cheers!

'I have seen the parents of some boys,' continued Squeers, turning over his papers, 'and they're so glad to hear how their sons are getting on, that there's no prospect at all of their sons' going away.—Which of course is a very pleasant thing to reflect upon, for all parties.'

Two or three hands went to two or three eyes, but the greater part of the young gentlemen—having no particular parents to speak of—were uninterested in the thing one way or other.

'I have had disappointments to contend against. Bolder's father was two pound ten short. Where is Bolder? Come here, Bolder.'

An unhealthy-looking boy, with warts all over his hands, stepped from his place to the master's desk, and raised his eyes to the face[1]; his own, quite white from the rapid beating of his heart.

'Bolder,' *speaking slowly, for he was considering, as the saying goes, where to have him.* 'Bolder, if your father thinks that because—why, what's this, sir?'

He caught up the boy's hand by the cuff of his jacket.

'What do you call this, sir?'

'I can't help the warts indeed, sir. They will come; it's the dirty work I think, sir—at least I don't know what it is, sir, but it's not my fault.'

'Bolder, you're an incorrigible young scoundrel, and as the last thrashing did you no good, we must see what another will do towards beating it out of you.'

Mr. Squeers fell upon the boy and caned him soundly.

'There; rub away as hard as you like, you won't rub that off in a hurry. Now let us see. A letter for Cobbey. Stand up, Cobbey.'

Another boy stood up, and eyed the letter very hard while Squeers made a mental abstract of it.[2]

'Oh! Cobbey's grandmother is dead, and his uncle John has took to drinking. Which is all the news his sister sends, *except eighteenpence, which will just pay for that broken square of glass. Mrs. Squeers, my dear, will you take the money?* Graymarsh, he's the next. Stand up, Graymarsh.'

Another boy stood up.

'Graymarsh's maternal aunt is very glad to hear he's so well and happy, and sends her respectful compliments to Mrs. Squeers, *and thinks she must be a angel.*[3] She likewise thinks Mr. Squeers is too good for this

[1] *sic:* 'Squeers's' deleted, 'the' inserted: but Dickens probably intended 'the master's'. Bolder replied, below ('I can't help the Warts...'), with 'Both Hands supplicatingly' (Wright).

[2] Here and during Graymarsh's aunt's letter, Squeers 'Reads from Book'; at 'Mrs. Squeers, ... will you take the money?' he 'Hands money' (Wright). The words from *except eighteenpence* are doubly underlined.

[3] *an angel* altered in *Berg* to *a angel*: but pronounced *a hangel* (Wright).

world; but hopes he may long be spared to carry on the business.[1] Would have sent the two pair of stockings as desired, but is short of money, so forwards a tract instead. Hopes, above all things, that Graymarsh will study to please Mr. and Mrs. Squeers, and look upon them *as his only friends*;[2] and that he will love Master Squeers; *and not object to sleeping five in a bed, which no Christian should.* Ah! a delightful letter. Very affecting indeed.'

It was[3] *affecting in one sense, for Graymarsh's maternal aunt was strongly supposed, by her more intimate friends, to be his maternal parent.*

'Mobbs's mother-in-law took to her bed on hearing that he wouldn't eat fat, and has been very ill ever since.[4] She wishes to know, by an early post, where he expects to go to, if he quarrels with his vittles; and with what feelings he *could*[5] turn up his nose at the cow's liver broth, after his good master had asked a blessing on it. This was told her in the London newspapers—not by Mr. Squeers, for he is too kind and too good to set anybody against anybody. Mobbs's mother-in-law is sorry to find Mobbs is discontented, which is sinful and horrid, and hopes Mr. Squeers will flog him into a happier state of mind; with this view, she has also stopped his halfpenny a week pocket-money, and given a double-bladed knife with a corkscrew in it, which she had bought on purpose for him, to the Missionaries.[6] A sulky state of feeling won't do. Cheerfulness and contentment must be kept up. Mobbs, come to me!'

The unhappy Mobbs moved slowly towards the desk, rubbing his eyes in anticipation of good cause for doing so; and soon afterwards retired by the side door, with as good cause as a boy need have.

Mr. Squeers then proceeded to open a miscellaneous collection of letters; some enclosing money, *which Mrs. Squeers 'took care of;'* and others referring to small articles of apparel, as caps and so forth, all of which the same lady stated to be too large, or too small, for anybody but young Squeers, *who would appear to have had most accommodating limbs, since everything that came into the school fitted him.*

In course of time, Squeers retired to his fireside, leaving Nicholas to take care of the boys in the school-room, which was very cold, and where a meal of bread and cheese was served out shortly after dark.

[1] Squeers repeated 'She likewise thinks Mr. Squeers is too good for this world' (Wright), and hereabouts, reflecting on this eulogy, he sagely observed that 'a good man struggling with his destiny is—a spectacle for things in general' (Field, p. 58); these words do not appear in Dickens's copies. Then, downcast by Graymarsh's aunt's having sent no money, his tone altered to a 'sulky sort of voice' (Wright).

[2] Doubly underlined, as is the next italicized phrase.

[3] *was* doubly underlined, the rest singly.

[4] For these words, Dickens substituted (according to Kate Field, p. 58)—'and has had a succession of cold and boiling water alternately running down her back ever since'.

[5] *could* italic in *Berg*.

[6] 'This is a very disagreeable letter' interpolated (Wright).

There was a small stove at that corner of the room which was nearest to the master's desk, and by it Nicholas sat down, depressed and self-degraded. * As he was absorbed in meditation he encountered the up-turned face of *Smike*, on his knees before the stove, picking a few cinders from the hearth and planting them on the fire. When he saw that he was observed, he shrunk back, expecting a blow.[1]

'You need not fear me. Are you cold?'[2]

'N-n-o.'

'You are shivering.'

'I am not cold. I am used to it.'

'Poor broken-spirited creature!'

If he had struck the wretched object, he would have slunk away without a word. But, now, he burst into tears.

'Oh, dear, oh dear! My heart will break. It will, it will.'

'Hush! Be a man; you are nearly one by years, God help you.'

'By years! Oh, dear, dear, how many of them! How many of them since I was a little child, younger than any that are here now! Where are they all!'

'Whom do you speak of?'

'My friends, myself—my—oh! what sufferings mine have been!'

'There is always hope.'

'No, no; none for me. Do you remember the boy that died here?'

'I was not here, you know, but what of him?'

'I was with him at night, and when it was all silent he cried no more for friends he wished to come and sit with him, but began to see faces round his bed that came from home; he said they smiled, and talked to him; and he died at last lifting his head to kiss them. What faces will smile on *me*[3] when *I* die here! Who will talk to *me* in those long long nights? They cannot come from home; they would frighten me, if they did, for I don't know what home is. Pain and fear, pain and fear for me, alive or dead. No hope, no hope!'[4]

The bell rang to bed; and the boy crept away. With a heavy heart Nicholas soon afterwards retired—no, not retired; there was no retirement there—followed—to the dirty and crowded dormitory.

[1] 'Hand to forehead' (Wright). Against the dialogue with Nicholas which follows, Wright has: 'Hands clasped as if in supplication', 'Sighs several times', 'Breathlessly'. It is unclear at which points Smike acted thus.

[2] This scene between Nicholas and Smike was 'painted with a vividness and pathos which told with thrilling effect upon the audience, and well deserved the burst of applause with which it was received' (*Cheltenham Examiner*, 8 January 1862).

[3] *me* . . . *I* . . . *me* all doubly underlined. 'Head from side to side' here (Wright).

[4] 'Oh!' with both hands up (Wright).

CHAPTER III[1]

MISS Fanny Squeers was in her three-and-twentieth year. If there be any grace of loveliness quite inseparable from that period of life, Miss Squeers must be presumed to have been possessed of it.[2] She was not tall like her mother, but short like her father—*from whom she inherited a remarkable expression of the right eye, something akin to having none at all.*

Miss Squeers had been spending a few days with a neighbouring friend, and had only just returned to the parental roof. * Questioning the servant, regarding the outward appearance and demeanour of Mr. Nickleby, the girl returned such enthusiastic replies, coupled with so many praises of his beautiful dark eyes, and his sweet smile, and his straight legs—*upon which last-named articles she laid particular stress; the general run of legs at Dotheboys Hall being crooked*—that Miss Squeers was not long in arriving at the conclusion that the new usher must be a very remarkable person, or, as she herself significantly phrased it, 'something quite out of the common.' And so Miss Squeers made up her mind that she would take a personal observation of Nicholas the very next day.

In pursuance of this design, the young lady watched the opportunity of her mother being engaged, and her father absent, and went *accidentally* into the school-room to get a pen mended: where, seeing nobody but Nicholas presiding over the boys, she blushed very deeply, and exhibited great confusion.

'I beg your pardon, I thought my Pa was—or might be—dear me, how very awkward!'

'Mr. Squeers is out.'

'Do you know will he be long, sir?'

'He said about an hour.'

'Thank you![3] I am very sorry I intruded, I am sure.' Miss Squeers said this, glancing from the pen in her hand, to Nicholas at his desk, and back again.

'If that is all you want,' said Nicholas, pointing to the pen, 'perhaps I can supply his place.'

Miss Squeers glanced at the door, as if dubious of the propriety of advancing nearer to a male stranger; then glanced round the school-room, as though in some measure reassured by the presence of forty boys; then sidled up to Nicholas, and delivered the pen into his hand.

'Shall it be a hard or a soft nib?'

[1] *III* altered to *2*, then back to *3*; cf. above, the footnote on the subtitle *4 Chapters*. Wright has a tantalizing note above this chapter-heading: 'Narration with forefinger and little finger'.

[2] Sentence deleted by Wright.

[3] '. . . that simper, lisp, and mien certainly belong to Mr. Squeers's offspring. Her "thank you" is perfect, and her . . . "As soft as possible, if you please" deserves to be perpetuated by a John Leech' (Field, p. 60).

'He *has*[1] a beautiful smile,' thought Miss Squeers. 'As soft as possible, if you please.'[2] Miss Squeers sighed. *It might be, to give Nicholas to understand that her heart was soft, and that the pen was wanted to match.*

Upon these instructions Nicholas made the pen; and when he gave it to Miss Squeers, Miss Squeers dropped it; and when he stooped to pick it up, Miss Squeers stooped too, and they knocked their heads together; whereat five-and-twenty little boys laughed: *being positively for the first and only time that half-year.* *

[3]Said Miss Squeers, as she walked away, 'I never saw such a pair of legs in the whole course of my life!'

In fact, Miss Squeers was in love with Nicholas Nickleby.

To account for the rapidity with which this young lady had conceived a passion for Nicholas, it may be necessary to state, that the friend from whom she had so recently returned, was a miller's daughter of only eighteen, who had engaged herself unto the son of a small corn-factor, resident in the nearest market town. Miss Squeers and the miller's daughter, being fast friends, had agreed together some two years before[4] (*according to a custom prevalent among young ladies*) that whoever was first engaged to be married, should straightway confide the mighty secret to the bosom of the other; in fulfilment of which pledge the miller's daughter, when her engagement was formed, came out express—*at eleven o'clock at night, as the corn-factor's son made an offer of his hand and heart at twenty-five minutes after ten by the Dutch clock in the kitchen*—and rushed into Miss Squeers's bed-room with the gratifying *intelligence*. Now Miss Squeers being five years older, had, since, been more than commonly anxious to return the compliment; but, either in consequence of finding it hard to please herself, or harder still to please anybody else, she had never had an opportunity so to do. The little interview with Nicholas had no sooner passed, than Miss Squeers, putting on her bonnet, made her way, with great precipitation, to her friend's house, and revealed how that she was—*not exactly engaged, but going to be—to a gentleman's son* (*none of your corn-factors*)—*to a gentleman's son*[5] of high descent, who had come down as teacher to Dotheboys Hall, under most mysterious and remarkable circumstances. Indeed, as Miss Squeers more than hinted, induced by the fame of her many charms, to seek her out, and woo and win her. *

'How I should like to see him!' exclaimed the friend.

'So you shall, 'Tilda; I should consider myself one of the most dis-

[1] *has* italic in *Berg* and in the novel.
[2] Marginal stage-direction *Sighing*.
[3] Marginal stage-direction *Legs*.
[4] *T & F* has 'ago' not 'before'.
[5] *not exactly engaged* and *none of your corn-factors* doubly underlined. Wright puts inverted commas round 'not exactly engaged, but going to be' and deletes 'to a gentleman's son (none of your corn-factors)'.

honorable creatures alive, if I denied you. Mother's going away tomorrow for two days to fetch some boys; and when she does, I'll ask you and your Intended, *John Browdie*,[1] up to tea, and have him to meet you.'[2]*

When such an opportunity occurred, it was Mr. Squeers's custom to drive over to the market town, every evening, on pretence of urgent business, and stop till ten or eleven o'clock at a tavern he much affected. As the contemplated party was not in his way, therefore, but rather afforded a means of compromise with Miss Squeers, he readily yielded his assent, and willingly told Nicholas that he was expected to take his tea in the parlour that evening, at five o'clock.

To be sure Miss Squeers was in a flutter, and to be sure she was dressed out to the best advantage: with her hair—she wore it in a crop—curled in five distinct rows,[3] up to the very top of her head, and arranged over the doubtful eye; to say nothing of the blue sash which floated down her back, or the worked apron, or the long gloves, or the scarf of green gauze, worn over one shoulder and under the other; or any of the numerous devices which were to be as so many arrows to the heart of Mr. Nickleby. She had scarcely completed these arrangements, when the friend arrived.†

The servant brought in the tea-things, and, soon afterwards, somebody tapped at the room door.

'There he is! Oh 'Tilda! I do so palpitate!'

'Hush! Hem! Say, come in.'

'Come in.'[4] And in walked Nicholas.

'Good evening,' *said that young gentleman, all unconscious of his conquest.* 'I understood from Mr. Squeers that I was expected?'

'Oh yes; it's all right. (Don't, 'Tilda!) Father don't tea with us, but you won't mind that, I dare say. We are only waiting for one more gentleman. ('Tilda, don't!).'[5] *

'Well,' thought Nicholas, 'as I am here, and seem expected to be amiable, it's of no use looking like a goose. I may as well accommodate myself to the company.' So he saluted Miss Squeers and the friend with gallantry, and drew a chair to the teatable, and began to make himself probably as much at home as ever an Usher was in his principal's parlor.

The ladies were in full delight at this, when the expected swain arrived

[1] Doubly underlined.

[2] *T & F* here has a paragraph (beginning 'It so fell out'), deleted in *Berg*. Wright deletes it in his copy of *T & F*, and correctly notes other such minor textual differences between *T & F* and *Berg*.

[3] 'Dramatic Both Hands' (Wright).

[4] Pronounced 'Coh/m/e' (Wright)—presumably Fanny's attempt at gentility.

[5] The bracketed asides to 'Tilda are late additions, not printed in *T & F*. Earlier, Dickens had interpolated Fanny's formal introduction 'Mr. Nickelby, 'Tilda; 'Tilda, Mr. Nickleby'—a fitting overture, wrote Kate Field, to 'one of the cleverest of *petite* comedies'. Dickens, she thought, was not at his best in this Reading until this scene (Field, pp. 59–60).

(with his hair damp from washing) in a clean shirt, whereof the collar might have belonged to some giant ancestor, and a white waistcoat of similar dimensions.

'Well, John,' said Miss Matilda Price (*which, by-the-by, was the name of the miller's daughter*).

'Weel,' said John *with a grin that even the collar could not conceal.*

'I beg your pardon,' interposed Miss Squeers, hastening to do the honours, 'Mr. Nickleby—Mr. John Browdie.'

'Servant, sir,' said John, who was about six feet six, with a face and body rather above the due proportion.

'Yours, sir,' replied Nicholas, *making fearful ravages on the bread and butter.*[1]

Mr. Browdie was not a gentleman of great conversational powers, so he grinned twice more, and having now bestowed his customary mark of recognition on every person in company, grinned at nothing particular, and helped himself to food.

'Old wooman awa', bean't she?'

Miss Squeers nodded assent.

Mr. Browdie gave a grin of special width, as if he thought that really was something to laugh at, and went to work at the bread and butter with vigour. It was quite a sight to behold how he and Nicholas emptied the plate between them.

'Ye wean't get bread and butther ev'ry neight, I expect, mun,' said Mr. Browdie, *after he had sat staring at Nicholas a long time over the empty plate.* 'Ecod, they dean't put too much intiv 'em. Ye'll be nowt but skeen and boans if you stop here long eneaf. Ho! ho! ho!'

'You are facetious, sir.'

'Na; I dean't know, but t'oother teacher, 'cod he wur a learn 'un, he wur.'

The recollection of the last teacher's leanness seemed to afford Mr. Browdie the most exquisite delight.

'I don't know whether your perceptions are quite keen enough, Mr. Browdie, to enable you to understand that your remarks are offensive, but they are——'

Miss Price stopped her admirer's mouth as he was about to answer.[2] 'If you say another word, John, I'll never forgive you, or speak to you again.'

[3]'Weel, my lass, I dean't care aboot 'un; let 'un gang on, let 'un gang on.'

It now became Miss Squeers's turn to intercede with Nicholas and the

[1] 'High laugh' (Wright)—John Browdie's, of course, which was a great feature of his characterization. His height is increased by six inches, in the Reading (above).

[2] 'Quick' (Wright). Miss Price is 'Quick' in other speeches, too, he notes.

[3] Marginal stage-direction *Snapping fingers.*

effect of the double intercession, was, that he and John Browdie shook hands across the table with much gravity. Such was the imposing nature of this ceremonial, that Miss Squeers shed tears.

'What's the matter, Fanny?' said Miss Price.

'Nothing, 'Tilda.'

'There never was any danger,' said Miss Price, 'was there, Mr. Nickleby?'

'None at all. Absurd.'

'Say something kind to her, and she'll soon come round. Here! Shall John and I go into the little kitchen, and come back presently?'

'Not on any account! What on earth should you do that for?'

'Well, you *are*[1] a one to keep company.'

'What do you mean? I am not a one to keep company at all. You don't mean to say that you think—'

'O no! I think nothing at all. Look at her, dressed so beautiful, and looking so well—really *almost* handsome. I am ashamed at you.'

'My dear girl, what have I got to do with her dressing beautifully or looking well?'

'Come, don't call me a dear girl!'[2] (*She smiled a little though, for she was pretty, and a coquette in her small way, and Nicholas was good-looking, and she supposed him the property of somebody else, which were all reasons why she should be gratified to think she had made an impression on him.*) 'Come; we're going to have a game at cards.' *She tripped away and rejoined the big Yorkshireman, and they sat down to play speculation.*

'There are only four of us, 'Tilda,' *said Miss Squeers, looking slyly at Nicholas*: 'so we had better go partners, two against two.'[3]

'What do you say, Mr. Nickleby?'

'With all the pleasure in life.' *And quite unconscious of his heinous offence, he 'went partners' with Miss Price.*

[4]'Mr. Browdie, shall we make a bank against them?'

The Yorkshireman assented—apparently quite overwhelmed by the usher's impudence—and Miss Squeers darted a spiteful look at her friend.

The deal fell to Nicholas, and the hand prospered.

'We intend to win everything.'

''Tilda *has*[5] won something she didn't expect, I think, haven't you, dear?'

'Only a dozen and eight, love.'

'How dull you are to-night!'

'No, indeed. I am in excellent spirits. I was thinking *you* seemed out of sorts.'

[1] *are*, like *almost* just below, italic in the novel and in *Berg*.
[2] 'Slappingly' (Wright).
[3] 'Back of fingers of left hand to mouth' (Wright).
[4] Marginal stage-direction *Spiteful*.
[5] *has*, like *you* just below, italic in the novel and in *Berg*.

'Me? Oh no!'

'Your hair's coming out of curl, dear.'

'Never mind me;[1] you had better attend to your partner, Miss.'

'Thank you for reminding her. So she had.'

[2]*The Yorkshireman flattened his nose, once or twice, with his clenched fist, as if to keep his hand in, till he had an opportunity of exercising it upon the nose of some other gentleman; and Miss Squeers tossed her head with such indignation, that the gust of wind raised by the multitudinous curls in motion, nearly blew the candle out.*

'I never had such luck, really,' exclaimed Miss Price, after another hand or two. 'It's all along of you, Mr. Nickleby, I think. I should like to have you for a partner always.'

'I wish you had.'

'You'll have a bad wife, though, if you always win at cards.'

'Not if your wish is gratified. I am sure I shall have a good one in that case.'

To see how Miss Squeers tossed her head, and [3]how the corn-factor flattened his nose, the while!

'We have all the talking to ourselves, it seems,' *said Nicholas, looking good-humouredly round the table as he took up the cards for a fresh deal.*

'You do it so well, that it would be a pity to interrupt, wouldn't it, Mr. Browdie?'

'Nay, we do it in default of having anybody else to talk to.'

'We'll talk to you, you know, if you'll say anything.'

'Thank you, 'Tilda.'

'Or you can talk to each other, if you don't choose to talk to us. John, why don't *you*[4] say something?'

'Say summat?'

'Ay, and not sit there so silent and glum.'

[5]'Weel, then! what I say's this—Dang my boans and boddy, if I stan' this ony longer. Do ye gang whoam wi' me; and do yon loight an' toight young whipster, look sharp out for a brokken head, next time he cums under my hond.'

'Mercy on us, what's all this?'

'Cum whoam, tell 'e, cum whoam!'

Here, Miss Squeers burst into tears; in part from vexation, and in part from an impotent desire to lacerate somebody's countenance with her fair finger-nails. *

[1] 'Snappishly' (Wright).

[2] Marginal stage-direction *Nose*. Dickens's enactment of this piece of business was delightful (Kent, p. 147).

[3] Marginal stage-direction *Nose* repeated. Browdie slapped his nose again at 'Say summat?' (Wright).

[4] *you* doubly underlined.

[5] Marginal stage-direction *Striking Table.*

'Why, and here's Fanny in tears now! What can be the matter?'

'Oh! you don't know, Miss, of course you don't know. Pray don't trouble yourself to inquire,' said Miss Squeers, *producing that change of countenance which children call, making a face.*

'Well, I'm sure!'

'And who cares whether you are sure or not, ma'am?'

'You are monstrous polite, ma'am.'

'I shall not come to you to take lessons in the art, ma'am!'

'Oh! you needn't take the trouble to make yourself plainer than you are, ma'am, however, because that's quite unnecessary.'

Miss Squeers, in reply, turned very red, and thanked God that she hadn't got the bold faces of some people. Miss Price, in rejoinder, congratulated herself upon not being possessed of the envious feeling of other people; whereupon Miss Squeers made some general remark touching the danger of associating with low persons.[1] *In which Miss Price entirely coincided.*

''Tilda, artful and designing 'Tilda! I wouldn't have a child named 'Tilda—not to save it from its grave!'

(*Here John Browdie, a little nettled*, wound up the evening by remarking, 'Weel, weel, weel! As to the matther o' thot, Fonny, it'll be time eneaf to think aboot neaming of it when it cooms.')[2]

CHAPTER IV[3]

THE poor creature, Smike, since the night Nicholas had spoken kindly to him in the school-room, had followed him to and fro, content only to be near him.[4] He would sit beside him for hours; and a word would brighten up his care-worn face, and call into it a passing gleam, even of happiness.

Upon this poor being, all the spleen and ill-humour that could not be vented on Nicholas were bestowed. It was no sooner observed that he had

[1] Wright puts inverted commas round 'low persons'. Fanny's next words, he notes, spoken 'Cryingly'.

[2] This brief exchange, from ch. 42, was printed in *Berg*, but slightly altered after *T & F* was printed. *Berg* contains one more printed page, to conclude Chapter III, but it is all deleted. *T & F*, however, retains one paragraph from this page, based upon the paragraph (near the end of ch. 9 of the novel) beginning 'They were no sooner gone ...' and ending with Fanny 'moaning in her pocket-handkerchief'. Wright deletes it.

[3] The 1858 printing began Chapter IV with 'The cold, feeble, dawn of a January day' (the opening of ch. 13 of the novel). Dickens then decided to amplify the text here, so with scissors and paste he stuck in some extracts from the end of ch. 12; these, subsequently much cut and amended, now precede 'The cold, feeble, dawn ...', but the paragraph which followed that in the novel ('It needed a quick eye ...'), printed in the 1858 copies, is deleted in both *Berg* and *Suzannet*. A slightly different, and shorter, version of the extracts from ch. 12 is written and pasted in to the *Suzannet* text at this point, to begin its Chapter III. Wright heads this chapter: 'Holding book over box'.

[4] Wright deletes from here until the end of the next paragraph ('... Smike paid for all').

become attached to Nicholas, than stripes and blows, morning, noon, and night, were his portion. Squeers was jealous of the influence his man had so soon acquired in the school, and the slighted Miss Squeers now hated Nicholas, and Mrs. Squeers hated him, and Smike paid for all.

One night the poor soul was poring hard over a book, vainly endeavouring to master some task which a child of nine years old could have conquered with ease, but which, to the brain of the crushed boy of nineteen, was a hopeless mystery.

Nicholas laid his hand upon his shoulder.

'I can't do it.'

'Do not try. * You will do better, poor fellow, when I am gone.'

'Gone! Are you going?'

'I cannot say.[1] I was speaking more to my own thoughts than to you. I shall be driven to that at last!' said Nicholas. 'The world is before me, after all.'

'Is the world as bad and dismal as this place?'

'Heaven forbid,'[2] replied Nicholas, pursuing the train of his own thoughts, 'it's hardest, coarsest toil, is happiness to this.'

[3]'Should I ever meet you there?'

'Yes,' willing to soothe him.

'No, no! Should I—should I— Say I should be sure to find you.'

'You would, and I would help you, and not bring fresh sorrow on you as I have done here.'

The boy caught both his hands and uttered a few broken sounds which were unintelligible. Squeers entered, at the moment, and he shrunk back into his old corner.

Two days later, the cold, feeble dawn of a January morning was stealing in at the windows of the common sleeping-room, when Nicholas, raising himself on his arm, looked among the prostrate forms in search of one.* †

'Now, then,' *cried Squeers, from the bottom of the stairs*, 'are you going to sleep all day, up there—'

'We shall be down directly, sir.'

'Down directly! Ah! you had better be down directly, or I'll be down upon some of you in less time than directly. Where's that Smike?'

Nicholas looked round again. *

'He is not here, sir.'

'Don't tell me a lie. He is.'

'He is not. Don't tell me one.'

Squeers bounced into the dormitory, and, swinging his cane in the air ready for a blow, darted into the corner where Smike usually lay at night. The cane descended harmlessly. There was nobody there.

[1] 'Hand to brow' (Wright).

[2] 'Hand uplifted' (Wright).

[3] Wright inserts: 'If I was to go into the world, should I . . .'

'What does this mean? Where have you hid him?'

'I have seen nothing of him, since last night.'

'Come, you won't save him this way. Where is he?'

'At the bottom of the nearest pond for anything I know.'

'D—n you, what do you mean, by that?' In a fright, Squeers inquired of the boys whether any one of them knew anything of their missing schoolmate.

There was a general hum of denial, in the midst of which, one shrill voice was heard to say (*as*, *indeed*, *everybody thought*[1]):—

'Please, sir, I think Smike's run away, sir.'

'Ha! who said that?'

Mr. Squeers made a plunge into the crowd, and caught a very little boy, the perplexed expression of whose countenance as he was brought forward, seemed to intimate that he was uncertain whether he was going to be punished or rewarded for his suggestion. He was not long in doubt.

'You think he has run away, do you, sir?'

'Yes, please sir.'

'And what reason have you to suppose that any boy would want to run away from this establishment? Eh?'

The child raised a dismal cry, by way of answer, and Mr. Squeers beat him until he rolled out of his hands. He mercifully allowed him to roll away.

'There! Now if any other boy thinks Smike has run away, I shall be glad to have a talk with him.'

Profound silence.[2]

'Well, Nickleby, *you*[3] think he has run away, I suppose?'

'I think it extremely likely.'

'Maybe you know he has run away?'

'I know nothing about it.'[4]

'He didn't tell you he was going, I suppose?'

'He did not. I am very glad he did not, for it would then have been my duty to have told you.'

'Which no doubt you would have been devilish sorry to do.'

'I should, indeed.'

Mrs. Squeers had listened to this conversation, from the bottom of the stairs; but, now losing all patience, she hastily made her way to the scene of action.

'What's all this here to do? What on earth are you talking to *him*[5] for,

[1] *thought* doubly underlined.

[2] Wright alters 'Profound' to 'Dead'.

[3] *you* italic in the novel and in *Berg*. Wright notes that, after 'Well, Nickleby', Dickens interpolated: 'you look to me'.

[4] 'Determinedly' (Wright).

[5] *him*, like *must* . . . *must* just below, doubly underlined.

Squeery! * The cow-house and stable are locked up, so Smike can't be there; and he's not down stairs anywhere, for the girl has looked. He *must* have gone York way, and by a public road. * He *must* beg his way, and he could do that nowheres but on the public road. Now, if you takes the chaise and goes one road, and I borrows Swallows's chaise, and goes t'other, what with keeping our eyes open, and asking questions, one or other of us is moral sure to lay hold of him.'

The lady's plan was put in execution without delay, * Nicholas remaining behind, in a tumult of feeling.[1] Death, from want and exposure, was the best that could be expected from the prolonged wandering of so helpless a creature, through a country of which he was ignorant. There was little, perhaps, to choose between this and a return to the tender mercies of the school: but Nicholas lingered on, in restless anxiety, until the evening of next day, when Squeers returned, alone.

'No news of the scamp!' *

Another day came, and Nicholas was scarcely awake when he heard the wheels of a chaise approaching the house. It stopped, and the voice of Mrs. Squeers was heard, ordering a glass of spirits for somebody—*which was in itself a sufficient sign that something extraordinary had happened.* Nicholas hardly dared to look out of window, but he did so, and the first object that met his eyes was wretched Smike: bedabbled with mud and rain, haggard and worn, and wild.

[2]'Lift him out,' said Squeers. 'Bring him in; bring him in!'

'Take care,' cried Mrs. Squeers. 'We tied his legs under the apron and made 'em fast to the chaise, to prevent his giving us the slip again.'

With hands trembling with delight, Squeers unloosened the cord; and Smike, more dead than alive, was brought in and locked up in a cellar, until such time as Mr. Squeers should deem it expedient to operate upon him. *

The news that the fugitive had been caught and brought back, ran like wild-fire through the hungry community, and expectation was on tiptoe all the morning. On tiptoe it remained until the afternoon; when Squeers, having refreshed himself with his dinner, and an extra libation or so, made his appearance (accompanied by his amiable partner) with a fearful instrument of flagellation, strong, supple, wax-ended, and new.

'Is every boy here?'[3]

Every boy was there, but every boy was afraid *to speak*; so Squeers glared along the lines to assure himself.[4]

[1] Wright deletes from here until the paragraph beginning 'Another day came'.

[2] Marginal stage-direction *Savage glee*; very dramatic, Wright noted.

[3] 'Hand down at side' (Wright).

[4] 'As Squeers was represented as "glaring along the lines"..., the Reader, instead of uttering one word of what the ruffianly schoolmaster ought then to have added: 'Each boy keep his place. Nickleby! You go to your desk, sir!" [speech deleted in *Berg*]

There was a curious expression in the usher's face; but he took his seat, without opening his lips in reply. Squeers left the room, and shortly afterwards returned, dragging Smike by the collar—*or rather by that fragment of his jacket which was nearest the place where his collar ought to have been.* *

'Now, what have you got to say for yourself? (*Stand a little out of the way, Mrs. Squeers, my dear; I've hardly got room enough.*)'[1]

'Spare me, sir!'

'Oh! that's all you've got to say, is it? Yes, I'll flog you within an inch of your life, and spare you that.'[2] *

One cruel blow had fallen on him, when Nicholas Nickleby cried, 'Stop!'[3]

'Who cried stop?'

'I did. This must not go on.'

'Must not go on!'

'Must not! Shall not! I will prevent it. You have disregarded all my quiet interference in this miserable lad's behalf; you have returned no answer to the letter I wrote you, in which I begged forgiveness for him, and offered to be responsible that he would remain quietly here. Don't blame me for this public interference. You have brought it upon yourself; not I.'

'Sit down, you beggar!'

'Touch him again at your peril! I will not stand by, and see it done. My blood is up, and I have the strength of ten such men as you. By Heaven I will not spare you, if you drive me on! I have personal insults to avenge, and my indignation is aggravated by the cruelties practised in this wicked den. Take care; for if you raise the devil in me, the consequences will fall heavily upon your head!'

Squeers, in a violent outbreak of wrath, spat at him, and struck him a blow across the face. Nicholas instantly sprang upon him, wrested his weapon from his hand, pinned him by the throat, and beat the ruffian till he roared for mercy. *

He flung him away with all the force he could muster, and the violence of his fall precipitated *Mrs. Squeers over an adjacent form*; *Squeers, striking*

—instead of uttering one syllable of this, contented himself with his own manuscript direction [in *Berg*], in one word—*Pointing*. The effect of this simple gesture was startling—particularly when, after the momentary hush with which it was always accompanied, he observed quietly,—"There was a curious expression in the usher's face; but he took his seat, without opening his lips in reply"' (Kent, p. 149).

[1] Doubly underlined; spoken with 'a horrible relish' (Kent, p. 150).

[2] 'Threateningly—clenched' (Wright).

[3] '"Stop!" was cried out in a voice that made the rafters ring—even the lofty rafters of St. James's Hall.' Squeers replied 'with the glare and snarl of a wild beast', gave another 'frightful look' on 'Must not go on!' and 'screamed out' his final speech, 'Sit down, you beggar!' (Kent, p. 150).

his head against the same form in his descent, lay at his full length on the ground, stunned and motionless.[1]

Having brought affairs to this happy termination, and having ascertained, to his satisfaction, that Squeers was only stunned, and not dead (*upon which point he had had some unpleasant doubts at first*), Nicholas packed up a few clothes in a small valise, and, finding that nobody offered to oppose his progress, marched boldly out by the front door, and struck into the road. Then such a cheer arose as the walls of Dotheboys Hall had never echoed before, and would never respond to again. When the sound had died away, the school was empty; and of the crowd of boys, *not one remained*.[2]

When Nicholas had cooled, sufficiently to give his present circumstances some reflection, they did not appear in an encouraging light; he had only four shillings and odd pence in his pocket, and was something more than two hundred and fifty miles from London.

Lifting up his eyes, he beheld a horseman coming towards him, discovered to be no other than Mr. John Browdie, carrying a thick ash stick.

'I am in no mood for more noise and riot, and yet, do what I will, I shall have an altercation with this honest blockhead, and perhaps a blow or two from yonder cudgel.'

There appeared reason to expect it, for John Browdie no sooner saw Nicholas, than he reined in his horse, and waited until such time as he should come up.

'Servant, young genelman.'

'Yours.'

'Weel; we ha' met at last.'

'Yes.—Come! We parted on no very good terms the last time we met; it was my fault; but I had no intention of offending you, and no idea that I was doing so. I was very sorry for it, afterwards. Will you shake hands?'

[1] The episode of Squeer's castigation was performed with a 'startling reality' (*Manchester Guardian*, 26 October 1868). Squeers's 'ruffianism . . . so much scandalized the audience . . . that, when the reader came to the fight . . ., there came a loud outburst of applause as [Squeers] fell senseless to the floor, just as if they had all been witnesses of an actual combat' (Edinburgh *Courant*, 19 April 1866). This reviewer noted, however, that 'One scarcely knows whether to take Mr. Squeers for a mere ruffian or a humorist in disguise.' The Reading, if not reconciling these two roles, fully exploited the dramatic possibilities of both.

[2] Underlined trebly. These two sentences from ch. 64 are pasted into *Berg*. They do not appear in *Suzannet*, which at this point makes a large deletion (the whole of the John Browdie episode). Its text jumps from Nicholas's leaving Dotheboys Hall ('. . . marched boldly out by the front door, and struck into the road') to the paragraph beginning 'He did not travel far . . .' From there to the end, the *Suzannet* text is virtually identical with *Berg*.

[1]'Shake honds! Ah! that I weel! But wa'at be the matther wi' thy feace, mun? It be all brokken loike.'

'It is a cut—a blow; but I returned it to the giver, and with good interest.'

'Noa, did'ee though? Well deane! I loike 'un for thot.'

'The fact is, I have been ill-treated.'

'Noa! Dean't say thot.'

'Yes, I have, by that man Squeers, and I have beaten him soundly, and am leaving this place in consequence.'

'What!' cried John Browdie, *with such an ecstatic shout, that the horse shyed at it.* 'Beatten the schoolmeasther! Ho! ho! ho! Beatten the schoolmeasther. Who ever heard o' the loike o' that noo! Giv' us thee hond agean, yoongster. Beatten the schoolmeasther! Dang it, I loove thee for't.'[2]

When his mirth had subsided, he inquired what Nicholas meant to do? On his replying, to go straight to London, he shook his head, and inquired if he knew how much the coaches charged, to carry passengers so far?

'No, I do not; but it is of no great consequence to me, for I intend walking.'

'Gang awa' to Lunnun afoot! (Stan' still, tellee, old horse.) Hoo much cash hast thee gotten?'

'Not much, but I can make it enough. Where there's a will, there's a way, you know.'

John Browdie pulled out an old purse, and insisted that Nicholas should borrow from him whatever he required.

'Dean't be afeard, mun, tak' eneaf to carry thee whoam. Thee'lt pay me yan day, a' warrant.'

Nicholas would by no means be prevailed upon to borrow more than a sovereign, with which loan Mr. Browdie was fain to content himself after many entreaties that he would accept of more. (*He observed, with a touch of Yorkshire caution, that if Nicholas didn't spend it all, he could put the surplus by, till he had an opportunity of remitting it carriage free*).

'Tak' that bit o' timber to help thee on wi', mun; keep a good heart, and bless thee. Beatten the schoolmeasther! 'Cod it's the best thing a've heerd this twonty year!'[3]

John set spurs to his horse, and went off at a smart canter. Nicholas watched the horse and rider until they disappeared over the brow of a distant hill, and then set forward on his journey.

He did not travel far, that afternoon, for by this time it was nearly dark;

[1] Marginal stage-direction *Shaking hands.* Wright deletes the *e* in 'shake'.

[2] After 'I loove thee for 't' Wright interpolates: 'Givens thee hand again'.

[3] 'Shaking hands closed or clasped together up and down and out and in from his [*word(s) lost*]' (Wright).

so, he lay, that night, at a cottage, where beds were let cheap; and, rising betimes next morning, made his way before night to Boroughbridge. There he stumbled on an empty barn; and in a warm corner stretched his weary limbs, and fell asleep.

When he awoke next morning, he sat up, rubbed his eyes, and stared at some motionless object in front of him.

'Strange! It cannot be real—and yet I—I am awake! Smike!'

It was Smike indeed.[1]

'Why do you kneel to me?'

'To go with you—anywhere—everywhere—to the world's end—to the churchyard. Let me go with you, oh do let me.[2] You are my home—my kind friend—take me with you, pray!'

He had followed Nicholas, it seemed; had never lost sight of him all the way; had watched while he slept, and when he halted for refreshment; and had feared to appear, sooner, lest he should be sent back.

'Poor fellow! Your hard fate denies you any friend but one, and he is nearly as poor and helpless as yourself.'

'May I—may I go with you? I will be your faithful hard-working servant. I want no clothes; these will do very well. I only want to be near you.'

'And you shall. And the world shall deal by you as it does by me, till one or both of us shall quit it for a better. [3] Come!'

So, he strapped his burden on his shoulders, and, taking his stick in one hand, extended the other to his delighted charge. And so they passed out of the old barn, together.[4]

THE END OF THE READING

[1] Wright deletes this sentence.

[2] 'Both hands supplicatingly' (Wright).

[3] 'Beckoningly'; below, at 'extended the other', 'Thrust out his hand' (Wright).

[4] 'I am inclined to suspect', Dickens wrote to Wilkie Collins after the first performance of *Nickleby*, 'that the impression of protection and hope derived from Nickleby's going away protecting Smike is exactly the impression—this is discovered by chance—that an Audience most likes to be left with' (*N*, iii. 248).

MR. BOB SAWYER'S PARTY

Bob, as Dickens usually called this item, was first performed on 30 December 1861 as an afterpiece to *Nickleby* and was an immediate and lasting success: it always provoked 'unflagging merriment', 'every line was a hit', to quote two critics. 'Dickens was as much born to read *Mr. Bob Sawyer's Party* as he was to create it', wrote Kate Field: this episode from *Pickwick* (ch. 32) became unforgettable, once his performance had been seen: 'What has struck you heretofore as a diamond no better than its fellows is magically transformed into a Kohinoor' (p. 92). This had never been one of the more legendary moments in the novel. Mr. Pickwick has little to say or do in it, Sam Weller is not present, and the two medical students (Bob Sawyer and Benjamin Allen) are relatively minor characters. In the Reading, indeed, the leading figure was Mr. Jack Hopkins, whose only appearance in the novel occurs in this chapter. Dickens's choice of the *Trial* episode from *Pickwick*, three years earlier, had been almost inevitable: as was noted above, this had long been a universal favourite in recitals and theatrical adaptations. Why Dickens chose Bob Sawyer's party, from a novel which contained so many better-loved episodes, can only be surmized, for he never discusses this Reading in his extant letters. It was of course a conveniently short episode, and complete in itself; also it provided opportunities for character-acting unlike those in the rest of the repertoire (notably the group of raffish and eventually tipsy young gentlemen). But, as such tributes as Kate Field's affirm, Dickens had been right in his instinct that the episode had great comic potential and that he could do it full justice.

To make every line a hit, Dickens had worked hard as arranger and reviser, as well as performer, sharpening and heightening the verbal comedy, simplifying the narrative, and eliminating irrelevances. Mr. Pickwick's companions (Tupman, Winkle and Snodgrass) are unnamed, and the final arrivals at the party, who in the novel had all been individually described, are summarized as 'the rest of the company . . . among whom there was . . . a sentimental young gentleman with a very nice sense of honour'. This is Mr. Noddy, later to figure in the tipsy quarrel—but his antagonist in the quarrel is changed. In the novel this had been Mr. Gunter, but Gunter is eliminated from the Reading, and his role goes to augment Jack Hopkins's. At one point, at least, there was an expansion of the script which does not appear in the prompt-copy (*Suzannet*); this was in Bob Sawyer's reproachful speech to Jack Hopkins about the over-boisterous 'chorusing' which leads to the abrupt termination of the party (see below, p. 206, note 1).

The most striking development in the text centres on Jack Hopkins. His abrupt idiom, present in the novel, is intensified. A good instance is his anecdote of the boy who swallowed his sister's beads, where Dickens deleted most of the definite and indefinite articles, and inserted the word 'necklace' five more times, to exploit the amusement caused by his pronouncing it 'neck-lass' or 'neck-luss'. According to Kate Field, 'the mere pronunciation of the word "necklace" inspired as much laughter as is usually accorded to a low-comedy man's best "point"', and she uses a quasi-musical notation to indicate how the Jack Hopkins voice—upon which so much of the Reading's success depended—was manipulated (pp. 97–9). Jack Hopkins's face, stance, and manner were much relished, as well as his voice. He stood, stiff-necked, with the inflated air of an incipient public man; his voice escaped as best it could 'between closed teeth, and from a mouth apparently full of mush'; his hands were thrust into his pantaloon-trousers, 'an attitude that, when accompanied by an oscillation of the body, as in Jack Hopkins's case, always indicates superior wisdom' (Kent, p. 158; Field, p. 96). He was one of Dickens's most comic platform creations, much enjoyed by Dickens himself, too. In the company of his Boston friends, James and Annie Fields, he 'laughed till the tears ran down his cheeks over Bob Sawyer's party and the remembrance of the laughter he had seen depicted on the faces of people the night before. Jack Hopkins was such a favorite with J[ames] that D. made up the face again and went over the necklace story until we roared aloud' (*Memories of a Hostess*, ed. M. A. DeWolfe Howe (1923), p. 146).

Bob was full of good 'faces' as well as comic lines and situations (Benjamin Allen's was another), and in Betsey the servant-girl it had the most striking of Dickens's platform creations; with only sixty words to speak, she was unforgettable, 'an incomparably comic character' (Field, p. 93). In *Bob*, wrote a Portsmouth journalist, 'Mr. Dickens's histrionic powers had full scope'. His rapid changes of face, manner, and tone of voice, from one character to another, were all 'admirably accomplished and intensely amusing', and the audience left the entertainment, which this item concluded, 'scarcely knowing which to admire most, the great and eminent novelist or the clever reciter'.

Mr. Bob Sawyer's Party

THERE is a repose about Lant Street in the borough of Southwark in the county of Surrey, which sheds a gentle melancholy upon the soul. A house in Lant Street would not come within the denomination of a first-rate residence, in the strict acceptation of the term; but if a man wished to abstract himself from the world—to remove himself from the reach of temptation—to place himself beyond the possibility of any inducement to look out of window—he should by all means go to Lant Street.

In this happy retreat are colonized a few clear-starchers, a sprinkling of journeymen bookbinders, one or two prison agents for the Insolvent Court, several small housekeepers who are employed in the Docks, a handful of milliners, and a seasoning of jobbing tailors. The majority of the inhabitants either direct their energies to the letting of furnished apartments, or devote themselves to the healthful pursuit of mangling. The chief features in the still life of the street, are green shutters, lodging-bills, brass door-plates, and bell-handles; the principal specimens of animated nature are the pot-boy, the muffin youth, and the baked-potato man. *The population is migratory, usually disappearing on the verge of quarter-day, and generally by night. Her Majesty's revenues are seldom collected in this happy valley; the receipt of rent is dubious; and the water communication is frequently cut off.*

Mr. Bob Sawyer[1] embellished one side of the fire, in his first-floor front, early on the evening for which he had invited Mr. Pickwick to a friendly party; and his chum *Mr. Ben Allen* embellished the other side. The preparations for the reception of visitors appeared to be completed. The umbrellas in the passage had been heaped into the little corner outside the back-parlour door; the bonnet and shawl of the landlady's servant had been removed from the banisters; there were not more than two pairs of pattens on the street-door mat; and a kitchen candle, with a long snuff, burnt cheerfully on the ledge of the staircase window. Mr. Bob Sawyer had himself purchased the spirits, and had returned home in attendance on the bearer to preclude the possibility of their being absconded with, or delivered at the wrong house. The bottles were ready in the bed-room; a little table had been got from the parlour to play at cards on; and the glasses of the establishment, together with those which had been borrowed for the occasion from the public-house, were all drawn up in a tray on the floor of the landing outside the door.

Notwithstanding the highly satisfactory nature of these arrangements,

[1] Like *Mr. Ben Allen* below, this is doubly underlined.

there was a cloud on the countenance of Mr. Bob Sawyer, as he sat by the fire, and there was a sympathising expression, too, in the features of Mr. Ben Allen, and[1] melancholy in his voice, as he said,

'Well, it *is*[2] unlucky that your landlady, Mrs. Raddle, should have taken it in her head to turn sour, just on this occasion. She might at least have waited till to-morrow.'

'That's her malevolence; that's her malevolence.[3] She says that if I can afford to give a party I ought to be able to afford to pay her confounded "little bill."'

'How long has it been running?'[4] *A bill, by the bye, is the most extraordinary locomotive engine that the genius of man ever produced. It would keep on running during the longest lifetime, without ever once stopping of its own accord.*

'Only a quarter, and a month or so.'

Ben Allen coughed and directed a searching look between the two top bars of the stove.

'It'll be a deuced unpleasant thing if she takes it into her head to let out, when those fellows are here, won't it?'

'Horrible, horrible.'

Here, a low tap was heard at the room door, and Mr. Bob Sawyer looked expressively at his friend, and bade the tapper come in; whereupon a dirty slipshod girl, in black cotton stockings, thrust in her head, and said,

'Please, Mister Sawyer, Missis Raddle wants to speak to *you*.'[5]

Before Mr. Bob Sawyer could return an answer, this young person suddenly disappeared with a jerk, *as if somebody had given her a violent pull behind*. This mysterious exit was no sooner accomplished, than there was another tap at the door.

Mr. Bob Sawyer glanced at his friend with a look of abject apprehension, and once more cried 'Come in.'

The permission was not at all necessary, for, before Mr. Bob Sawyer had uttered the words, a little fierce woman bounced into the room, all in a tremble with passion, and pale with rage.

[6]'Now Mr. Sawyer, if you'll have the kindness to settle that little bill

[1] 'and' is supplied by *T & F*: omitted in *Suzannet*.

[2] *is* italic in *Suzannet* and in the novel. [3] 'Sneering' (Wright).

[4] 'Sucking thumb and in a sneaking manner.' Bob Sawyer's reply ('and a month or so')—'Pulling down waistcoat' (Wright).

[5] *you* italic in *Suzannet* and in the novel. 'Slow', notes Wright. Charles Kent, like the critics quoted in the headnote, was dazzled by the impersonation of Betsey: 'No one had ever realized the crass stupidity of that remarkable young person . . . until her first introduction in these Readings, with "Please, Mister Sawyer, Missis Raddle wants to speak to *you*!"—the dull, dead-level of her voice ending on the last monosyllable with a series of inflections amounting to a chromatic passage' (p. 33).

[6] 'Very quick' (Wright).

of mine I'll thank you, because I've got my rent to pay this afternoon, and my landlord's a waiting below now.' *Here the little woman rubbed her hands, and looked steadily over Mr. Bob Sawyer's head, at the wall behind him.*

'I am very sorry to put you to any inconvenience, Mrs. Raddle, but—'

'Oh, it isn't any inconvenience. I didn't want it particular before to-day; leastways, as it has to go to my landlord directly, it was as well for you to keep it, as me. You promised me this afternoon, Mr. Sawyer, and every gentleman as has ever lived here, has kept his word, sir, as of course anybody as calls himself a gentleman, do.' *Mrs. Raddle tossed her head, bit her lips, rubbed her hands harder, and looked at the wall more steadily than ever.*

'I am very sorry, Mrs. Raddle, but the fact is, that I have been disappointed in the City to-day.'—*Extraordinary place that city. Astonishing number of men always getting disappointed there.*

'Well, Mr. Sawyer, and what is that to me, sir?'

'I—I—have no doubt, Mrs. Raddle,' *said Bob, blinking this last question*, 'that before the middle of next week we shall be able to set ourselves quite square, and go on, on a better system, afterwards.'

This was all Mrs. Raddle wanted. She had bustled up to the apartment of the unlucky Bob, so bent upon going into a passion, that, in all probability, payment would have rather disappointed her. She was in excellent order for a little relaxation of the kind: having just exchanged a few introductory compliments with Mr. Raddle in the front kitchen.

'Do you suppose, Mr. Sawyer,' *elevating her voice for the information of the neighbours*, 'do you suppose that I'm a-going day after day to let a fellar occupy my lodgings as never thinks of paying his rent, nor even the very money laid out for the fresh butter and lump sugar that's bought for his breakfast, nor the very milk that's took in, at the street door? Do you suppose as a hard-working and industrious woman which has lived in this street for twenty year (ten year over the way and nine year and three quarter in this very house) has nothing else to do, but to work herself to death after a parcel of lazy idle fellars, that are always smoking and drinking, and lounging, when they ought to be glad to turn their hands to anything that would help 'em to pay their bills?'

'My good soul,' *interposed Mr. Benjamin Allen.*

'Have the goodness to keep your observashuns to yourself, sir, I beg,' *suddenly arresting the rapid torrent of her speech, and addressing the third party with impressive slowness and solemnity.* 'I am not aweer, sir, that you have any right to address your conversation to *me*.[1] I don't think I let these apartments to *you*, sir.'

'No, you certainly did not.'

'Very good, sir. Then p'r'aps, sir, as a medical studient, you'll confine

[1] *me* doubly underlined, as is *you* in the next sentence.

yourself to breaking the arms and legs of the poor people in the hospitals, and will keep yourself *to*[1] yourself, sir, or there may be some persons here as will make you, sir.'

'But you are such an unreasonable woman.'

'I beg your parding, young man. But will you have the goodness to call me that again, sir?'[2]

'I didn't make use of the word in any invidious sense, ma'am.'

'I beg your parding, young man. But who do you call a woman? Did you make that remark to me, sir?'

'Why, bless my heart!'

'Did you apply that name to me, I ask of you, sir?'—*with intense ferocity, and throwing the door wide open.*

'Why of course I did.'

'Yes of course you did,' *backing gradually to the door, and raising her voice for the special behoof of Mr. Raddle in the kitchen.* 'Yes, of course you did! And everybody knows that you may safely insult me in my own ouse while my husband sits sleeping down stairs, and taking no more notice than if I was a dog in the streets. He ought to be ashamed of himself (*sob*) to allow his wife to be treated in this way by a parcel of young cutters and carvers of live people's bodies, that disgraces the lodgings (*another sob*), and leaving her exposed to all manner of abuse; a base faint-hearted, timorous wretch, that's afraid to come up stairs, and face the ruffinly creatures—that's afraid—that's afraid to come!' *Mrs. Raddle paused to listen whether the repetition of the taunt had roused her better half; and, finding that it had not been successful, proceeded to descend the stairs with sobs innumerable: when there came a loud double knock at the street door. Hereupon she burst into a fit of weeping, which was prolonged until the knock had been repeated six times, when, in an uncontrollable burst of mental agony, she threw down all the umbrellas, and disappeared into the back parlour.*

'Does Mr. Sawyer live here?' said Mr. Pickwick,[3] when the door was opened.

'Yes, first floor. It's the door straight afore you, when you gets to the top of the stairs.' *Having given this instruction, the handmaid, who had been brought up among the aboriginal inhabitants of Southwark, disappeared, with the candle in her hand, down the kitchen stairs.*

Mr. Pickwick and his two friends stumbled up stairs, where they were

[1] *to* italic in *Suxannet* and in the novel.

[2] Here 'Mrs. Raddle's anger rose through an increasing *crescendo*.' Ben Allen replied 'meekly and somewhat uneasy on his own account,' and her response came 'louder and more imperatively' (Kent, pp. 155–6).

[3] Wright has 'Idiotic voice' in the margin here: but probably the description was meant to refer to Betsey, who has the next speech, delivered as a 'lugubrious and monotonously intoned response, all on one note' (Kent, p. 156).

received by the wretched Bob, *who had been afraid to go down, lest he should be waylaid by Mrs. Raddle.*

'How are you? Glad to see you,—take care of the glasses.' *This caution was addressed to Mr. Pickwick, who had put his foot in the tray.*

'Dear me, I beg your pardon.'

'Don't mention it, don't mention it. I'm rather confined for room here, but you must put up with all that, when you come to see a young bachelor. Walk in. You've seen Mr. Ben Allen before, I think?' *Mr. Pickwick shook hands with Mr. Benjamin Allen, and his friends followed his example.* They had scarcely taken their seats when there was another double knock.

'I hope that's Jack Hopkins! Hush. Yes, it is. Come up, Jack; come up.'

A heavy footstep was heard upon the stairs, and Jack Hopkins presented himself.[1]

'You're late, Jack?'

'Been detained at Bartholomew's.'

'Anything new?'

'No, nothing particular.[2] Rather a good accident brought into the casualty ward.'

'What was that, sir?'

'Only a man fallen out of a four pair of stairs' window; but it's a very fair case—very fair case indeed.'

'Do you mean that the patient is in a fair way to recover?'

'No. No, I should rather say he wouldn't. There must be a splendid operation though, to-morrow—magnificent sight if Slasher does it.'

'You consider Mr. Slasher a good operator?'

'Best alive. Took a boy's leg out of the socket last week—boy ate five apples and a gingerbread cake—exactly two minutes after it was all over, boy said he wouldn't lie there to be made game of; and he'd tell his mother if they didn't begin.'

'Dear me!'

'Pooh! that's nothing. Is it, Bob?'

'Nothing at all.'

'By the bye, Bob,' said Hopkins, *with a scarcely perceptible glance at Mr. Pickwick's attentive face*, 'we had a curious accident last night. A child was brought in, who had swallowed a necklace.'[3]

'Swallowed what, sir?'

'A necklace. Not all at once, you know, that would be too much—*you*

[1] On Jack Hopkins's physical presence and voice, see headnote.
[2] 'Hands in pocket' (Wright).
[3] On the delivery of this anecdote, see headnote. Kent reproduces (facing his p. 152) the page from *Suzannet* showing Dickens's deletion of 'a' and 'the' and his insertion of the joke-word 'necklace'. In Jack Hopkins's next speech, *you* is italic in *Suzannet* and in the novel. Some punctuation in this speech is supplied from *T & F.*

couldn't swallow that if the child did—eh, Mr. Pickwick, ha! ha! No, the way was this;—child's parents, poor people, lived in a court. Child's eldest sister bought a necklace,—common necklace—large black wooden beads. Child, being fond of toys, cribbed necklace, hid necklace, played with necklace, cut string of necklace, and swallowed a bead. Child thought it capital fun, went back next day, and swallowed another bead.'

'Bless my heart, what a dreadful thing! I beg your pardon, sir. Go on.'

'Next day, child swallowed two beads; day after that, treated himself to three beads—so on, till in a week's time he had got through the necklace—five-and-twenty beads. Sister—industrious girl, seldom treated herself to bit of finery—cried eyes out, at loss of necklace; looked high and low for necklace; but, I needn't say, didn't find necklace. Few days afterwards, family at dinner—baked shoulder of mutton, and potatoes—child wasn't hungry, playing about the room, when family suddenly heard devil of a noise, like small hail storm. "Don't do that, my boy," said father. "I ain't a doin' nothing," said child. "Well, don't do it again," said father. Short silence, and then noise worse than ever. "If you don't mind what I say, my boy," said father, "you'll find yourself in bed, in something less than a pig's whisper." Gave child a shake to make him obedient, and such a rattling ensued as nobody ever heard before. "Why, damme, it's *in*[1] the child!" said father, "he's got the croup in the wrong place!" "No, I haven't, father," said child, beginning to cry, "it's the necklace; I swallowed it, father."—Father caught child up, and ran with him to hospital: beads in boy's stomach rattling all the way with the jolting; and people looking up in the air, and down in the cellars, to see where unusual sound came from. He's in the hospital now, and makes such a devil of noise when he walks about, that they're obliged to muffle him in a watchman's coat, for fear he should wake the patients!' †

Here another knock at the door announced the rest of the company—five in number—among whom there was, as presently appeared, a sentimental young gentleman with a very nice sense of honor. The little table was wheeled out; the bottles were brought in, and the succeeding three hours were devoted to a round game at sixpence a dozen. †

When the last deal had been declared, and the profit and loss account of fish and sixpences adjusted, to the satisfaction of all parties, Mr. Bob Sawyer rang for supper, and the visitors squeezed themselves into corners while it was getting ready.

It was not so easily got ready as some people may imagine. First of all, it was necessary to awaken the girl who had fallen asleep with her face on the kitchen table; this took time, and even when she did answer the bell, another quarter of an hour was consumed in fruitless endeavours to impart to her a distant glimmering of reason. The man to whom the

[1] *in* italic in *Suzannet* and in the novel.

order for the oysters had been sent, had not been told to open them; it is a very difficult thing to open an oyster with a limp knife or a two-pronged fork; and very little was done in this way. Very little of the beef was done either; and the ham (which was also from the German-sausage shop round the corner) was in a similar predicament. However, there was plenty of porter in a tin can; and the cheese went a great way, for it was very strong.

After supper, more bottles were put upon the table, together with a paper of cigars.[1] Then there was an awful pause; and this awful pause was occasioned by an embarrassing occurrence.

The fact is, the girl was washing the glasses. The establishment boasted four; which is not mentioned to its disparagement, for there never was a lodging-house yet, that was not short of glasses. The establishment's glasses were little thin feeble tumblers, and those which had been borrowed from the public-house were great dropsical, bloated articles, each supported on a huge gouty leg. This would have been in itself sufficient to have possessed the company with the real state of affairs; even if the young person of all work had not prevented the possibility of any misconception arising in the mind of any gentleman upon the subject, by forcibly dragging every man's glass away, long before he had finished his beer, and audibly stating, despite the winks of Mr. Bob Sawyer, that it was to be conveyed down stairs, and washed forthwith.

It is an ill wind that blows nobody any good. The prim man in the cloth boots,[2] who had been unsuccessfully attempting to make a joke during the whole time the round game lasted, saw his opportunity and seized it. The instant the glasses disappeared, he commenced a long story 'about a great public character, whose name I have forgotten, making a particularly happy reply to another eminent and illustrious individual whom I have never been able to identify.' He enlarged with great minuteness upon divers collateral circumstances, distantly connected with the anecdote in hand, but said, 'For the life of me, I cannot recollect at this precise moment what the anecdote is,[3] although I have been in the habit of telling the story with great applause for the last ten years. Dear me, it is a very extraordinary circumstance.'

'I am sorry you have forgotten it,' *said Mr. Bob Sawyer, glancing eagerly*

[1] A long optional cut is here indicated in *Suzannet*, by marginal lines; it ends at the close of the paragraph (four paragraphs below) ending '... the very best story he had ever heard'.

[2] Unmentioned before now, because the brief description of him and his companions (in the novel paragraph beginning 'Another knock at the door') had been deleted; Dickens had substituted the brief phrase about 'the rest of the company—five in number ...' A rare example of Dickens's having failed to make his cuts consistent.

[3] *Suzannet* has 'at that precise moment what the anecdote was ...'; *T & F* rightly corrects this to 'this ... is ...', in line with Dickens's substitution here of direct speech for indirect.

at the door, as he thought he heard the noise of glasses jingling—'very sorry.'

'So am I, because I know it would have afforded so much amusement. Never mind; I dare say I shall manage to recollect it, in the course of half an hour or so.'

The prim man arrived at this point, just as the glasses came back, when Mr. Bob Sawyer, who had been absorbed in attention, said *he should very much like to hear the end of it*, for, *so far as it went*, it was, without exception, the very best story he had ever heard.

The sight of the tumblers restored Bob to a degree of equanimity he had not possessed since his interview with his landlady. His face brightened up, and he began to feel quite convivial.

'Now, Betsey,' *dispersing the tumultuous little mob of glasses the girl had collected in the centre of the table*; 'now, Betsey, the warm water: be brisk, there's a good girl.'

'You can't have no warm water.'

'No warm water!'

'No, Missis Raddle said you warn't to have none.'

'Bring up the warm water instantly—instantly!'

'No, I can't. Missis Raddle raked out the kitchen fire afore she went to bed, and locked up the kittle.'

'Never mind: never mind. Pray don't disturb yourself about such a trifle,' *said Mr. Pickwick, observing the conflict of Bob Sawyer's passions, as depicted in his countenance*, 'cold water will do very well.'[1]

'My landlady is subject to some slight attacks of mental derangement. I fear I must give her warning.'

'No, don't.'

'I fear I must. Yes, I'll pay her what I owe her, and give her warning to-morrow morning.' *Poor fellow! how devoutly he wished he could!*

Mr. Bob Sawyer's attempts to rally under this last blow communicated a dispiriting influence to the company, the greater part of whom, with the view of raising their spirits, attached themselves with extra cordiality to the cold brandy and water. The first effects of these libations were displayed in an outbreak of hostilities between the youth with the nice sense of honor and Mr. Hopkins. At last the youth with the nice sense of honor felt it necessary to come to an understanding on the matter; when the following clear understanding took place.

'Sawyer.'

'Well, Noddy.'

'I should be very sorry, Sawyer, to create any unpleasantness at any friend's table, and much less at yours, Sawyer,—very; but I must take this opportunity of informing Mr. Hopkins that he is no gentleman.'

[1] Another optional cut is indicated here, in *Suzannet*, by marginal lines—from 'My landlady is subject . . .' to '*how devoutly he wished he could!*'

'And *I*[1] should be very sorry, Sawyer, to create any disturbance in the street in which you reside, but I'm afraid I shall be under the necessity of alarming the neighbours by pitching the person who has just spoken, out o' window.'

'I should like to see you do it, sir.'

'You shall *feel*[2] me do it in half a minute, sir.'

'I request that you'll favour me with your card, sir.'

'I'll do nothing of the kind, sir.'

'Why not, sir?'

'Because you'll stick it up over your chimney-piece, and delude your visitors into the false belief that a gentleman has been to see you, sir.'

'Sir, a friend of mine shall wait on you in the morning.'

'Sir, I am very much obliged to you for the caution, and I'll leave particular directions with the servant to lock up the spoons.'

At this point the remainder of the guests interposed, and remonstrated with both parties on the impropriety of their conduct. A vast quantity of talking ensued, in the course of which Mr. Noddy gradually allowed his feelings to overpower him, and professed that he had ever entertained a devoted personal attachment towards Mr. Hopkins. To this, Mr. Hopkins replied that on the whole he preferred Mr. Noddy to his own mother; on hearing this admission, Mr. Noddy magnanimously rose from his seat, and proffered his hand to Mr. Hopkins. Mr. Hopkins grasped it; and everybody said the whole dispute had been conducted in a manner[3] which was highly honourable to both parties concerned.

'And now, just to set us going again, Bob, I don't mind singing a song.' *Hopkins, incited by applause, plunged at once into* '*The King, God bless him,*' *which he sang as loud as he could, to a novel air, compounded of the* '*Bay of Biscay*' *and* '*A Frog he would a wooing go.*' *The chorus was the essence of the song; and, as every gentleman sang it to the tune he knew best, the effect was very striking.*

It was at the end of the chorus to the first verse, that Mr. Pickwick held up his hand in a listening attitude, and said, as soon as silence was restored,

'Hush! I beg your pardon. I thought I heard somebody calling from up stairs.'

A profound silence ensued; and Mr. Bob Sawyer was observed to turn pale.

'I think I hear it now. Have the goodness to open the door.'

The door was no sooner opened than all doubt on the subject was removed by a voice screaming from the two-pair landing, 'Mr. Sawyer! Mr. Sawyer!'

[1] *I* italic in *Suzannet* and in the novel.

[2] *feel* italic in *Suzannet* and in the novel.

[3] 'Here he would sometimes gag' (Kent, p. 160).

'(It's my landlady. I thought you were making too much noise.[1]—) Yes, Mrs. Raddle.'

'What do you mean by this, Mr. Sawyer? Aint it enough to be swindled out of one's rent, and money lent out of pocket besides, and insulted by your friends that dares to call themselves men, without having the house turned out of window, and noise enough made to bring the fire-engines here at two o'clock in the morning?—Turn them wretches away.'

'You ought to be ashamed of yourselves,' *said the voice of Mr. Raddle, which appeared to proceed from beneath some distant bed-clothes.*[2]

'Ashamed of themselves! Why don't you go down and knock 'em every one down stairs? You would if you was a man.'

'I should if I was a dozen men, my dear, but they've the advantage of me in numbers, my dear.'

'Ugh, you coward! *Do*[3] you mean to turn them wretches out, Mr. Sawyer?'

'They're going, Mrs. Raddle, they're going. (I am afraid you'd better go. I *thought*[4] you were making too much noise.)—They're only looking for their hats, Mrs. Raddle; they are going directly.'

Mrs. Raddle, thrusting her night-cap over the banisters just as Mr. Pickwick[5] emerged from the sitting-room. 'Going! What did they ever come for?'

'My dear ma'am'—remonstrated Mr. Pickwick, looking up.

'Get along with you, you old wretch!' said Mrs. Raddle, *hastily withdrawing the night-cap.* 'Old enough to be his grandfather, you villin! You're worse than any of 'em.'

Mr. Pickwick found it in vain to protest his innocence, so hurried down stairs into the street, closely followed by the rest.* The visitors having all departed, in compliance with this rather pressing request of Mrs. Raddle, the luckless Mr. Bob Sawyer was left alone, to meditate on the probable events of the morrow, and the pleasures of the evening.

THE END OF MR. BOB SAWYER'S PARTY

[1] 'Bob . . . turned reproachfully on the over-boisterous Jack Hopkins, with, "I *thought* you were making too much noise, Jack. You're such a fellow for chorusing! You're always at it. You came into the world chorusing; and I believe you'll go out of it chorusing"' (Kent, p. 161). Kate Field gives a slightly different version of this unwritten interpolation: Bob tells Jack that it is all his fault, 'because he will sing chorus—that he was born chorus-y, lives chorus-y, and will die chorus-y' (p. 95).

[2] The words *which appeared . . . distant bed-clothes* are underlined doubly in *Suzannet*. 'From an artistic point of view, the cleverest portions [of the Reading] were those in which the servant girl appeared . . . and the solitary ejaculation of Mr. Raddle from the recesses of the bed clothes' (*Yorkshire Post*, 17 April 1869).

[3] *Do* italic in *Suzannet* and in the novel.

[4] *thought* italic in *Suzannet* and in the novel.

[5] The underlining in *Suzannet* ends prematurely, at the end of a line.

DOCTOR MARIGOLD

THE market cheap-jack, Doctor Marigold, had enormously delighted readers of the *All the Year Round* Christmas number for 1865, *Doctor Marigold's Prescriptions*, where his monologues occupied the first and last chapters. Most reviewers agreed with *The Times* (6 December 1865) that Marigold was 'a masterly sketch, and one that deserves a place in our memories beside the best picture ever drawn by Dickens'. Dickens was due to resume his career as a public reader—which had been virtually in abeyance since 1862—in April 1866, when he began the first of his series under the management of Messrs. Chappell: so, as a new attraction, he devised *Doctor Marigold*. 'I have got him up with immense pains,' he told Forster on 11 March 1866 (*Life*, p. 701), and a week later, having now rehearsed the new piece over 200 times, he gave a highly successful trial reading to a few friends, including Browning, Forster, Wilkie Collins, Charles Kent, and the actor Charles Fechter.

Marigold (paired with *Bob Sawyer*) opened the new series, at St. James's Hall, London, on 10 April 1866, and it immediately became, as it remained, one of the most popular items in the repertoire. As George Dolby recalled (p. 9), it 'more than realized the anticipations of even the most sanguine of Mr Dickens's friends, whilst the public, and those who in various ways were more immediately interested in the Readings, were convinced that up to that time they had had but a very faint conception of Mr Dickens's power either as an adapter or an elocutionist'. His skill as an adapter was not in fact severely tested by this item. No rearrangement of episodes was necessary, only some abridgement. The prompt-copy (*Berg*) contains fewer deletions and performance-signs than usual; probably another copy had been used for those 200 rehearsals. An unusual feature is that underlining is used almost entirely to distinguish subdued or pathetic passages from comic ones. There are no marginal stage-directions, but happily W. M. Wright, in his copious notes on this Reading, provides plenty. The performance, he says, took about an hour and was prefaced by these words: 'I am to have the pleasure of reading to you tonight Dr. Marigold, whom I will leave to tell you his story in his own way.' The pronunciations which he records confirm that Marigold was a Dickensian Cockney: 'opposite' as 'oppo-sight', 'put' as in 'putty', 'joints' as 'jints', 'owner' as 'howner', 'rather' as 'rayther', 'waistcoat' as 'vaistcoat'.

Mid-Victorian readers and audiences loved such a blending of humour and pathos as *Marigold* offered. To adopt the critical vocabulary of the *Staffordshire Sentinel* (4 May 1867), the opening moments of this Reading quickly had the audience 'primed for cacchinatory exercise', but soon 'Marigold's pathetic description of his child's death rendered humid the majority of the eyes in the room'. For many people, Little Sophy's dying moments, while her father had to maintain his comic cheap-jack patter, constituted one of the most memorable passages in all the Readings. But *Marigold* was also 'one of the most humorous things he ever wrote', and Dickens's impersonation of the cheap-jack was one of his most admired performances: 'perhaps there is no character in which the great novelist appears to greater advantage' (*The Times*, 7 October 1868). As many observers noted, he *became* the cheap-jack—not only in the comic mimicry, but also in the rich sense he conveyed of the man's heart. His voices for Mim and Pickleson were also much enjoyed.

As in some of his other minor pieces, however, he has here been much more successful in imagining a character than in devising a story worthy of him. The *Manchester Guardian* critic put this point well: 'The cheap-jack himself is delightful, and was as irresistible as ever . . . But the story is sickly, and the episode of the deaf and dumb lover courting his deaf and dumb mistress is simply painful' (4 February 1867). Clearly, much depended on Dickens's performance and his audience. As a close friend, Annie Fields, thought, this was the 'subtlest of all the Readings' and required more of the listener than any other. This remark was provoked by one of his first American performances of this item, when the audience was not responsive, but Annie Fields and her husband 'were penetrated by it' (*James T. Fields: Biographical and Personal Sketches* (1881), p. 156). Generally, however, *Marigold* was very popular in America as well as Britain, despite these technical difficulties and the accent and the market-place patter being somewhat alien. Dickens described his first American performance of it as 'really a tremendous hit. The people doubted it at first, having evidently not the least idea what could be done with it, and broke out at last into a perfect chorus of delight. At the end they made a great shout, and gave a rush towards the platform as if they were going to carry me off' (*Life*, p. 777).

Doctor Marigold

CHAPTER I

I am a Cheap Jack,[1] and my own father's name was Willum Marigold. It was in his lifetime supposed by some that his name was William, but my own father always consistently said, No, it was Willum. On which point I content myself with looking at the argument this way:—If a man is not allowed to know his own name in a free country, how much is he allowed to know in a land of slavery?

I was born on the Queen's highway, but it was the King's at that time. A doctor was fetched to my own mother by my own father, when it took place on a common; and in consequence of his being a very kind gentleman, and accepting no fee but a tea-tray, I was named Doctor, out of gratitude and compliment to him. There you have me. Doctor Marigold.*

The doctor having accepted a tea-tray, you'll guess that my father was a Cheap Jack before me. You are right. He was. * And my father was a lovely one in his time at the Cheap Jack work. Now, I'll tell you what. I mean to go down into my grave declaring, that of all the callings ill-used in Great Britain, the Cheap Jack calling is the worst used. Why ain't we a profession? Why ain't we endowed with privileges? Why are we forced to take out a hawker's licence, when no such thing is expected of the political hawkers? Where's the difference betwixt us? Except that we are Cheap Jacks and they are Dear Jacks, I don't see any difference but what's in our favour.

For look here! Say it's election-time. I am on the footboard of my cart, in the market-place on a Saturday night. I put up a general miscellaneous lot. I say: 'Now here my free and independent woters, I'm a going to give you such a chance as you never had in all your born days, nor yet the days preceding. Now I'll show you what I am a going to do with you. Here's a pair of razors that'll shave you closer than the Board of Guardians; here's a flat-iron worth its weight in gold; here's a frying-pan artificially flavoured with essence of beefsteaks to that degree that you've only got for the rest of your lives to fry bread and dripping in it and there you are replete with animal food; here's a genuine chronometer watch, in such a solid silver case that you may knock at the door with it when you come home late from a social meeting, and rouse your wife and family and save up your knocker for the postman; and here's

[1] 'Shakes head determinedly' (Wright). He also noted that Dickens opened the monologue: 'I am known all over England.'

half a dozen dinner plates that you may play the cymbals with to charm the baby when it's fractious. Stop! I'll throw you in another article, and I'll give you that, and it's a rolling-pin, and if the baby can only get it well into its mouth when its teeth is coming, and rub the gums once with it, they'll come through double, in a fit of laughter equal to being tickled. Stop again! I'll throw you in another article, because I don't like the looks of you, for you haven't the appearance of buyers unless I lose by you, and because I'd rather lose than not take money to-night, and that article's a looking-glass in which you may see how ugly you look when you don't bid. What do you say now? Come! Do you say a pound? Not you, for you haven't got it. Do you say ten shillings? Not you, for you owe more to the tallyman. Well then, I'll tell you what I'll do with you. I'll heap 'em all on the footboard of the cart—there they are! razors, flat-iron, frying-pan, chronometer watch, dinner-plates, rolling-pin, and looking-glass—take 'em all away for four shillings, and I'll give you sixpence for your trouble!' This is me, the Cheap Jack.

But on the Monday morning, in the same market-place, comes the Dear Jack on the hustings—*his* cart—and what does *he* say?[1] 'Now my free and independent woters, I am a going to give you such a chance' (he begins just like me) 'as you never had in all your born days, and that's the chance of sending Myself to Parliament. Now I'll tell you what I am a going to do for you. Here's the interests of this magnificent town promoted above all the rest of the civilized and uncivilized earth. Here's your railways carried, and your neighbours' railways jockeyed. Here's all your sons in the Post-office. Here's Britannia smiling on you. Here's the eyes of Europe on you. Here's uniwersal prosperity for you, repletion of animal food, golden cornfields, gladsome homesteads, and rounds of applause from your own hearts, all in one lot, and that's myself. Will you take me as I stand?[2] You won't? Well, then, I'll tell you what I'll do with you. Come now! I'll throw you in anything you ask for. There! Church-rates, abolition of church-rates, more malt-tax, no malt-tax, uniwersal education to the highest mark or uniwersal ignorance to the lowest, total abolition of flogging in the army or a dozen for every private once a month all round, Wrongs of Men or Rights of Women—only say which it shall be, take 'em or leave 'em, and I'm of your opinion altogether, and the lot's your own on your own terms. There! You won't take it yet?[3] Well then, I'll tell you what I'll do with

[1] *his* and *he* italic in *All the Year Round* (*AYR*) and in *Berg*. Wright notes that Dickens used 'Both Hands' at the beginning of the Dear Jack's speech, spreading them out at 'Here's the interests of this magnificent town . . .'—*T & F* (but not *AYR* nor *Berg*) begins a new paragraph at 'But on the Monday morning': a creditably authorized clarification.

[2] 'Pause' (Wright).

[3] 'Pause and sigh and lean forward with' [other words cut away] (Wright).

you. Come![1] You *are* such free and independent woters, and I *am* so proud of you—you *are* such a noble and enlightened constituency, and I *am* so ambitious of the honour and dignity of being your member, which is by far the highest level to which the wings of the human mind can soar—that I'll tell you what I'll do with you. I'll throw you in all the public-houses in your magnificent town, for nothing. Will that content you? It won't? You won't take the lot yet? Well then, before I put the horse in and drive away, and make the offer to the next most magnificent town that can be discovered, I'll tell you what I'll do. Take the lot, and I'll drop two thousand pound in the streets of your magnificent town for them to pick up that can. Not enough? Now look here. This is the very furthest that I'm a going to. I'll make it two thousand five hundred. And still you won't? Here, missis! Put the horse—no, stop half a moment, I shouldn't like to turn my back upon you neither for a trifle, I'll make it two thousand seven hundred and fifty pound. There! Take the lot on your own terms, and I'll count out two thousand seven hundred and fifty pound on the footboard of the cart, to be dropped in the streets of your magnificent town for them to pick up that can. What do you say? Come now! You won't do better, and you may do worse. You take it? Hooray! Sold again, and got the seat!' *

I courted my wife from the footboard of the cart. I did indeed. She was a Suffolk young woman, and it was in Ipswich market-place right opposite the corn-chandler's shop. I had noticed her up at a window last Saturday that was, appreciating highly; I had took to her; and I had said to myself, 'If not already disposed of, I'll have that lot.' Next Saturday that come, I pitched the cart on the same pitch, and I was in very high feather indeed, keeping 'em laughing the whole of the time and getting off the goods briskly. At last I took out of my waistcoat pocket a small lot wrapped in soft paper, and I put it this way (looking up at the window where she was). 'Now here my blooming English maidens is a article, the last article of the present evening's sale, which I offer to only you the lovely Suffolk Dumplings biling over with beauty, and I won't take a bid of a thousand pound for, from any man alive. Now what is it? Why, I'll tell you what it is. It's made of fine gold, and it's not broke though there's a hole in the middle of it, and it's stronger than any fetter that ever was forged, though it's smaller than any finger in my set of ten. Why ten? Because when my parents made over my property to me, I tell you true, there was twelve sheets, twelve towels, twelve table-cloths, twelve knives, twelve forks, twelve table-spoons, and twelve tea-spoons, but my set of fingers was two short of a dozen and could never since be matched. Now what else is it? Come I'll tell you. It's a hoop of solid gold, wrapped in a silver curl-paper that I myself

[1] Italics in the following sentence printed in *AYR* and *Berg*.

took off the shining locks of the ever beautiful old lady in Threadneedle-street, London city. I wouldn't tell you so if I hadn't the paper to show, or you mightn't believe it even of me. Now what else is it? It's a man-trap and a handcuff, the parish stocks and a leg-lock, all in gold and all in one. Now what else is it? It's a wedding ring. Now I'll tell you what I'm a going to do with it. I'm not a going to offer this lot for money, but I mean to give it to the next of you beauties that laughs, and I'll pay her a visit to-morrow morning at exactly half after nine o'clock as the chimes go, and I'll take her out for a walk to put up the banns.' *She*[1] laughed, and got the ring handed up to her. When I called in the morning, she says 'Oh dear! It's never you, and you never mean it?' 'It's ever me,' says I, 'and I'm ever yours, and I ever mean it.' So we got married, after being put up three times—which, by-the-by, is quite in the Cheap Jack way again, and shows once more how the Cheap Jack customs pervade society.

She wasn't a bad wife, but she had a temper. If she could have parted with that one article at a sacrifice, I wouldn't have swopped her away in exchange for any other woman in England. Not that I ever did swop her away, for we lived together till she died, and that was thirteen year. Now my lords and ladies and gentlefolks all, I'll let you into a secret, though you won't believe it. Thirteen year of temper in a Palace would try the worst of you, but thirteen year of temper in a Cart would try the best of you. You are kept so very close to it in a cart, you see. There's thousands of couples among you, getting on like sweet ile upon a whet-stone, in houses five and six pairs of stairs high, that would go to the Divorce Court in a cart. Whether the jolting makes it worse, I don't undertake to decide, but in a cart it does come home to you and stick to you. Wiolence in a cart is *so* wiolent, and aggrawation in a cart is *so* aggrawating.[2]

We might have had such a pleasant life! A roomy cart, with the large goods hung outside and the bed slung underneath it when on the road, an iron pot and a kettle, a fireplace for the cold weather, a chimney for the smoke, a hanging shelf and a cupboard, a dog, and a horse. What more do you want? You draw off upon a bit of turf in a green lane or by the roadside, you hobble your old horse and turn him grazing, you light your fire upon the ashes of the last visitors, you cook your stew, and you wouldn't call the Emperor of France your father. But have a temper in the cart, flinging language and the hardest goods in stock at you, and where are you then? Put a name to your feelings.

[1] *She* italic in *AYR* and *Berg*.

[2] *so* . . . *so* italic in *AYR* and *Berg*. This description of temper in a cart 'fairly convulsed the audience with laughter' (*Chester Chronicle*, 26 January 1867). Wright has, at the top of the page containing the next paragraph: 'Rather nasal in tone—quiet way throughout—low tone'.

[1] My dog knew as well when she was on the turn as I did. Before she broke out, he would give a howl, and bolt.[2] How he knew it, was a mystery to me, but the sure and certain knowledge of it would wake him up out of his soundest sleep, and he would give a howl, and bolt. At such times I wished I was him.

The worst of it was we had a daughter born to us, and I love children with all my heart.[3] When she was in her furies she beat the child. This got to be so shocking, as the child got to be four or five year old, that I have many a time gone on with my whip over my shoulder, at the old horse's head, sobbing and crying worse than ever little Sophy did. For how could I prevent it? Such a thing is not to be tried with such a temper —in a cart—without coming to a fight. It's in the natural size and formation of a cart to bring it to a fight. And then the poor child got worse terrified than before, as well as worse hurt generally, and her mother made complaints to the next people we lighted on, and the word went round, 'Here's a wretch of a Cheap Jack been a beating his wife.'[4]

Little Sophy was such a brave child! She grew to be quite devoted to her poor father, though he could do so little to help her. She had a wonderful quantity of shining dark hair, all curling natural about her. It is quite astonishing to me now, that I didn't go tearing mad when I used to see her run from her mother before the cart, and her mother catch her by her hair, and pull her down by it, and beat her.

Yet in other respects her mother took great care of her. Her clothes were always clean and neat, and her mother was never tired of working at 'em. Such is the inconsistency of things. Our being down in the marsh country in unhealthy weather, I consider the cause of Sophy's taking bad low fever; but however she took it, once she got it she turned away from her mother for evermore, and nothing would persuade her to be touched by her mother's hand. She would shiver and say 'No, no, no,'[5] when it was offered at, and would hide her face on my shoulder, and hold me tighter round the neck.

The Cheap Jack business had been worse than ever I had known it, what with one thing and what with another (and not least what with railroads, which will cut it all to pieces, I expect, at last), and I was run

[1] 'Right hand uplifted' (Wright). At the bottom of the page containing the preceding and following paragraphs, Wright notes: 'Hands together working fingers'.

[2] 'And what an important character Doctor Marigold's dog becomes from just one or two references to his extraordinary sagacity! . . . The tone of the "howl" and action of the "bolt" are unutterably expressive' (Field, p. 77).

[3] 'When the good Doctor clasps his hands and presses them to his breast [here], as if he were embracing that pretty daughter . . ., you feel as if he really *did* love children. Moreover, you feel morally certain that Dickens loves children too' (Field, p. 78).

[4] 'Points' (Wright). The next paragraph was spoken 'Feelingly', and Dickens began it by interjecting 'My little Sophy! Ah!' (Wright).

[5] 'Sophy's voice' (Wright).

dry of money. For which reason, one night at that period of little Sophy's being so bad, either we must have come to a deadlock for victuals and drink, or I must have pitched the cart as I did.

I couldn't get the dear child to lie down or leave go of me, and indeed I hadn't the heart to try; so I stepped out on the footboard with her holding round my neck.[1] They all set up a laugh when they see us, and one chuckle-headed Joskin (that I hated for it) made the bidding, 'tuppence for her!'

'Now, you country boobies,' says I, *feeling as if my heart was a heavy weight at the end of a broken sash-line*,[2] † 'now let's know what you want to-night, and you shall have it. But first of all, shall I tell you why I have got this little girl round my neck? You don't want to know? Then you shall. She belongs to the Fairies. She is a fortune-teller. She can tell me all about you in a whisper, and can put me up to whether you're a going to buy a lot or leave it. Now do you want a saw? No, she says you don't, because you're too clumsy to use one. Your well-known awkwardness would make it manslaughter. Now I am a going to ask her what you do want. (*Then I whispered*, '*Your head burns so, that I am afraid it hurts you bad, my pet?*' *and she answered, without opening her heavy eyes*, '*Just a little, father.*') Oh! This little fortune-teller says it's a memorandum-book you want.[3] Then why didn't you mention it? Here it is. Look at it. Two hundred superfine hot-pressed wire-wove pages, ready ruled for your expenses, an everlastingly-pointed pencil to put 'em down with, a double-bladed penknife to scratch 'em out with, a book of printed tables to calculate your income with, and a camp-stool to sit down upon while you give your mind to it! Stop! And an umbrella to keep the moon off when you give your mind to it on a pitch dark night. Now I won't ask you how much for the lot, but how little? How little are you thinking of? Don't be ashamed to mention it, because my fortune-teller knows already. (*Then making believe to whisper, I kissed her, and she kissed me.*) Why, she says you are thinking of as little as three and threepence! I couldn't have believed it, even of you, unless she told me. Three and threepence! And a set of printed tables in the lot that'll calculate your income up to forty thousand a year! With an income of forty thousand a year, you grudge three and sixpence. Well then, I'll tell you my opinion. I so despise the threepence, that I'd sooner take three shillings. There. For three shillings, three shillings, three shillings! Gone. Hand 'em over to the lucky man.'

[1] 'Left Hand turned over on left breast as if holding Sophy. Smoothing back of same occasionally with Right Hand' (Wright).

[2] Like most of the underlining in this Reading, this is double; but as double underlining is not here being used to distinguish a different degree of emphasis (etc.) from single, this will not be noted henceforward.

[3] 'Right Hand slapping forehead and then throwing book on table' (Wright).

As there had been no bid at all, everybody looked about and grinned at everybody, while I touched little Sophy's face and asked her if she felt faint or giddy. 'Not very, father. It will soon be over.' Then turning from the pretty patient eyes, which were opened now, and seeing nothing but grins across my lighted grease-pot, I went on again in my Cheap Jack style. 'Where's the butcher?' (*My sorrowful eye had just caught sight of a fat young butcher on the outside of the crowd.*) 'She says the good luck is the butcher's. Where is he?' Everybody handed on the blushing butcher to the front, and there was a roar, and the butcher felt himself obliged to put his hand in his pocket and take the lot. The party so picked out, in general does feel obliged to take the lot. Then we had another lot the counterpart of that one, and sold it sixpence cheaper, which is always wery much enjoyed. So I went on in my Cheap Jack style till we had the ladies' lot—the tea-pot, tea-caddy, glass sugar basin, half a dozen spoons, and caudle-cup—*and all the time I was making similar excuses to give a look or two and say a word or two to my poor child. It was while the second ladies' lot was holding 'em enchained that I felt her lift herself a little on my shoulder, to look across the dark street. 'What troubles you, darling?' 'Nothing troubles me, father. I am not at all troubled. But don't I see a pretty churchyard over there?' 'Yes, my dear.' 'Kiss me twice, dear father, and lay me down to rest upon that churchyard grass so soft and green.'*[1] I staggers back into the cart with her head dropped on my shoulder, and I says to her mother, 'Quick. Shut the door! Don't let those laughing people see!' 'What's the matter?' she cries. 'O, woman, woman,' I tells her, 'you'll never catch my little Sophy by her hair again, for she's dead and flown away from you!'[2]

Maybe those were harder words than I meant 'em, but from that time forth my wife took to brooding, and would sit in the cart or walk beside it, hours at a stretch, with her arms crossed and her eyes looking

[1] 'Drop head forward'; Marigold's next speech given in a 'Hoarse voice' (Wright).

[2] As was remarked in the headnote, this episode of Sophy's death was the great moment of this Reading—indeed, one of the most celebrated moments in the repertoire. Superlatives were used liberally to describe Dickens's conception and performance of the scene: 'an irresistibly tragic power' (*Belfast Newsletter*, 21 May 1867); 'nothing could surpass in true refinement Mr. Dickens's style of rendering' it (*Manchester Guardian*, 4 February 1867); 'Incomparably the finest portion of all this wonderfully original sketch of Doctor Marigold, both in the Writing and in the Reading, . . . it was one that, by voice and look and manner, he himself most exquisitely delineated' (Kent, pp. 250–1). To conceive of such a scene showed 'certainly a surprising kind of dramatic genius', and Dickens's performance was 'worthy of the great French school of acting' (*Graphic*, 12 February 1870, p. 250). Apart from the minority view which found it sentimental clap-trap, only one technical criticism seems to have been made. Kate Field thought that Dickens made Marigold's bitter speech to his wife too loud; wanting to conceal his grief from the crowd, he would surely have 'muffled the cry of his heart. Did Doctor Marigold shout as Dickens does, he would alarm the entire neighbourhood' (p. 79).

on the ground.[1] When her furies took her (which was rather seldomer than before) they took her in a new way, and she banged herself about to the extent that I was forced to hold her. She got none the better for a little drink now and then. So sad our lives went on, till one summer evening, when as we were coming into Exeter out of the further West of England, we saw a woman beating a child in a cruel manner, who screamed, 'Don't beat me! O mother, mother, mother!'[2] *Then my wife stopped her ears and ran away like a wild thing, and next day she was found in the river.*

Me and my dog was all the company left in the cart now, and the dog learned to give a short bark when they wouldn't bid, and to give another and a nod of his head when I asked him: 'Who said half-a-crown? Are you the gentleman, sir, that offered half-a-crown?' He attained to an immense heighth of popularity, and I shall always believe taught himself entirely out of his own head to growl at any person in the crowd that bid as low as sixpence. But he got to be well on in years, and one night when I was conwulsing York with the spectacles, he took a conwulsion on his own account upon the very footboard by me, and it finished him.

Being naturally of a tender turn, I had dreadful lonely feelings on me arter this. I conquered 'em at selling times, having a reputation to keep (*not to mention keeping myself*), but they got me down in private and rolled upon me.

It was under those circumstances that I come acquainted with a giant. I might have been too high to fall into conversation with him, had it not been for my lonely feelings. For the general rule is, going round the country, to draw the line at dressing up. When a man can't trust his getting a living to his undisguised abilities, you consider him below your sort.[3] And this giant when on view figured as a Roman.

He was a languid young man, which I attribute to the distance betwixt his extremities. He had a little head and less in it, he had weak eyes and weak knees, and altogether you couldn't look at him without feeling that there was greatly too much of him both for his joints and his mind. But he was an amiable though timid young man (his mother let him out and spent the money), and we come acquainted when he was walking to ease the horse betwixt two fairs. He was called Rinaldo di Velasco, his name being Pickleson.

This giant otherwise Pickleson mentioned to me under the seal of confidence, that beyond his being a burden to himself, his life was made

[1] Marigold's compunction here, after his reproaches to his wife, 'was most affecting; and the story of the poor woman's brooding over her sorrow, and her death, . . . was given with a power that must have reached the heart of every one present' (*Birmingham Gazette*, 11 May 1866).

[2] 'Both Hands shaking to keep off the mother' (Wright).

[3] 'Sneeringly' (Wright). In the next sentence, and throughout whenever 'Roman' is mentioned, Dickens made it 'a hancient Roman' (Wright; Field, p. 79).

a burden to him, by the cruelty of his master towards a step-daughter who was deaf and dumb. Her mother was dead, and she had no living soul to take her part, and was used most hard. She travelled with his master's caravan, only because there was nowhere to leave her, and this giant otherwise Pickleson did go so far as to believe that his master often tried to lose her. He was such a very[1] languid young man, that I don't know how long it didn't take him to get this story out, but it passed through his defective circulation to his top extremity in course of time.

When I heard this account from the giant otherwise Pickleson, and likewise that the poor girl had beautiful long dark hair, and was often pulled down by it and beaten, I couldn't see the giant through what stood in my eyes. Having wiped 'em,[2] I give him sixpence (for he was kept as short as he was long), and he laid it out in two three-pennorths of gin-and-water, which so brisked him up, that he sang the Favourite Comic of Shivery Shakey, ain't it cold? A popular effect which his master had tried every other means to get out of him as a Roman, wholly in vain.

His master's name was Mim; a wery hoarse man, and I knew him to speak to. I went to that Fair as a mere civilian, leaving the cart outside the town, and I looked about the back of the Vans while the performing was going on, and at last sitting dozing against a muddy cart-wheel, I come upon the poor girl who was deaf and dumb.[3] At the first look I might almost have judged that she had escaped from the Wild Beast Show, but at the second I thought better of her, *and thought that if she was more cared for and more kindly used she would be like my child. She was just the same age that my own daughter would have been, if her pretty head had not fell down upon my shoulder[4] that unfortunate night.*

To cut it short, I spoke confidential to Mim while he was beating the gong outside betwixt two lots of Pickleson's publics, and I put it to him, 'She lies heavy on your own hands; what'll you take for her?' Mim was a most ferocious swearer. Suppressing that part of his reply, which was much the longest part, his reply was, 'A pair of braces.' 'Now I'll tell you,' says I, 'what I'm a going to do with you. I'm a going to fetch you half a dozen pair of the primest braces in the cart, and then to take her away with me.' Says Mim (again ferocious), 'I'll believe it when I've got the goods, and no sooner.' I made all the haste I could, lest he should think twice of it, and the bargain was completed: which Pickleson he was thereby so relieved in his mind that he come out at

[1] 'I don't know how Dickens does it, . . . but the complete vacuity of his face as he produces the word "wery", and the languor which accompanies his delivery of this sentence, absolutely makes you limp in joints and mind as Rinaldo di Velasco himself' (Field, p. 80). Kent noted that Dickens used a high falsetto voice for this giant (p. 250); it was, said Wright, 'like voice of man with fat chin in *Carol*'.

[2] 'Wipe eyes with fingers' (Wright). [3] 'Right Hand uplifted' (Wright).

[4] 'Looking up' (Wright).

his little back door, longways like a serpent, and give us Shivery Shakey in a whisper among the wheels at parting.

It was happy days for both of us when Sophy and me began to travel in the cart. *I at once give her the name of Sophy, to put her ever towards me in the attitude of my own daughter.* We soon made out to begin to understand one another, through the goodness of the Heavens, when she knowed that I meant true and kind by her. In a very little time she was wonderful fond of me. You have no idea what it is to have any body wonderful fond of you, unless you have been got down and rolled upon by the lonely feelings that I have mentioned as having once got the better of me.

You'd have laughed—or the rewerse—it's according to your disposition—if you could have seen me trying to teach Sophy. At first I was helped—you'd never guess by what—milestones. I got some large alphabets in a box, all the letters separate on bits of bone, and say we was going to WINDSOR,[1] I give her those letters in that order, and then at every milestone I showed her those same letters in that same order again, and pointed towards the abode of royalty. Another time I give her CART,[2] and then chalked the same upon the cart. Another time I give her DOCTOR MARIGOLD, and hung a corresponding inscription outside my waistcoat. People that met us might stare a bit and laugh, but what did *I* care if she caught the idea? She caught it after long patience and trouble, and then we did begin to get on swimmingly, I believe you! At first she was a little given to consider me the cart, and the cart the abode of royalty, but that soon wore off.

We had our signs, too, and they was hundreds in number.[3] Sometimes, she would sit looking at me and considering hard how to communicate with me about something fresh—how to ask me what she wanted explained—*and then she was (or I thought she was; what does it signify?) so like my child with those years added to her, that I half believed it was herself, trying to tell me where she had been to up in the skies, and what she had seen since that unhappy night when she flied away.* She had a pretty face, and now that there was no one to drag at her bright dark hair, and it was all in order, there was a something touching in her looks that made the cart most peaceful and most quiet, though not at all melancolly.

The way she learnt to understand any look of mine was truly surprising.

[1] Capitals here, and below for CART and DOCTOR MARIGOLD, printed thus in *AYR* and *Berg*. Below, *I* italic in *AYR* and *Berg*.

[2] 'Spells letters along table'—C/A/R/T, etc. (Wright).

[3] 'Feelingly' (Wright)—to the end of this paragraph. Significantly, Dickens deleted the uneasy little joke which (in *AYR*, and printed in *Berg*) ended the paragraph—'[N.B. In the Cheap Jack patter we generally sound it, lemonjolly, and it gets a laugh.]'

When I sold of a night, she would sit in the cart unseen by them outside,[1] and would give a eager look into my eyes when I looked in, and would hand me straight the precise article or articles I wanted. And then she would clap her hands and laugh for joy. And as for me, seeing her so bright, and remembering what she was when I first lighted on her, starved and beaten and ragged, leaning asleep against the muddy cart-wheel, it give me such heart that I gained a greater heighth of reputation than ever, and I put Pickleson down (by the name of Mim's Travelling Giant otherwise Pickleson) for a fypunnote in my will.

This happiness went on in the cart till she was sixteen year old.[2] By which time I began to feel not satisfied that I had done my whole duty by her, and to consider that she ought to have better teaching than I could give her. It drew a many tears on both sides when I commenced explaining my views to her, but what's right is right, and you can't neither by tears nor laughter do away with its character.

So I took her hand in mine, and I went with her one day to the Deaf and Dumb Establishment in London, and when the gentleman come to speak to us, I says to him: 'Now I'll tell you what I'll do with you sir. I am nothing but a Cheap Jack, but of late years I have laid by for a rainy day notwithstanding. This is my only daughter (adopted), and you can't produce a deafer nor yet a dumber. Teach her the most that can be taught her, in the shortest separation that can be named—state the figure for it—and I am game to put the money down. I won't bate you a single farthing sir, but I'll put down the money here and now, and I'll thankfully throw you in a pound to take it. There!' The gentleman smiled, and then, 'Well, well,' says he, 'I must first know what she has learnt already. How do you communicate with her?' Then I showed him, and she wrote in printed writing many names of things and so forth, and we held some sprightly conversation, Sophy and me, about a little story in a book which the gentleman showed her and which she was able to read. 'This is very extraordinary,' says the gentleman; 'is it possible that you have been her only teacher?' 'I have been her only teacher, sir,' I says, 'besides[3] herself.' 'Then,' says the gentleman, and more acceptable words was never spoke to me, 'you're a clever fellow, and a good fellow.' This he makes known to Sophy, who kisses his hands, claps her own, and laughs and cries upon it.

We saw the gentleman four times in all, and when he took down my name, and asked how in the world it ever chanced to be Doctor, it come out that he was own nephew by the sister's side, if you'll believe me,

[1] '[Word illegible] forefinger to lips left and across breast'; two sentences later, 'Clap hands' (Wright).

[2] 'Feelingly'—to the end of the paragraph; next paragraph 'Quick' (Wright).

[3] '... besides her bright—beautiful—clever self (Kiss Right Hand to Sophy 3 times between [words])' (Wright).

to the very Doctor that I was called after. This made our footing still easier, and he says to me:

'Now Marigold, tell me what more do you want your adopted daughter to know?'

'I want her sir to be cut off from the world as little as can be, considering her deprivations, and therefore to be able to read whatever is wrote, with perfect ease and pleasure.'

'My good fellow,' urges the gentleman, opening his eyes wide, 'why *I*[1] can't do that myself!'

I took his joke and give him a laugh (knowing by experience how flat you fall without it), and I mended my words accordingly.

'What do you mean to do with her afterwards?' asks the gentleman, with a sort of a doubtful eye. 'To take her about the country?'

'In the cart sir, but only in the cart.[2] She will live a private life, you understand, in the cart. I should never think of bringing her infirmities before the public. I wouldn't make a show of her for any money.'

The gentleman nodded and seemed to approve.

'Well,' says he, 'can you part with her for two years?'

'To do her that good—yes, sir.'

'There's another question,' says the gentleman, looking towards her: 'Can she part with you for two years?'

I don't know that it was a harder matter of itself (for the other was hard enough to me), but it was harder to get over.[3] However, she was pacified to it at last, and the separation betwixt us was settled. How it cut up both of us when it took place, and when I left her at the door in the dark of an evening, I don't tell. But I know this:—remembering that night, I shall never pass that same establishment without a heart-ache and a swelling in the throat; and I couldn't put you up the best of lots in sight of it with my usual spirit—no, not for five hundred pound reward from the Secretary of State for the Home Department, and throw in the honour of putting my legs under his mahogany arterwards.

Still, the loneliness that followed in the cart was not the old loneliness, because there was a term put to it however long to look forward to, and because I could think, when I was anyways down, that she belonged to me and I belonged to her. Always planning for her coming back, I bought in a few months' time another cart, and what do you think I planned to do with it? I'll tell you. I planned to fit it up with shelves, and books for her reading, and to have a seat in it where I could sit and see her read, and think that I had been her first teacher. Not hurrying over the job, I had the fittings knocked together in contriving ways

[1] *I* italic in *AYR* and *Berg*.

[2] 'Rather crossly' (Wright).

[3] Wright heads the page which begins here: 'Action with Both Hands throughout'. Against the rest of this paragraph, he writes: 'Monotonous low voice'.

under my own inspection, and here was her bed in a berth with curtains, and there was her reading-table, and here was her writing-desk, and elsewhere was her books, in rows upon rows, picters and no picters, bindings and no bindings, gilt-edged and plain, just as I could pick 'em up for her in lots up and down the country, North and South and West and East, Winds liked best and winds liked least, Here and there and gone astray, Over the hills and far away. And when I had got together pretty well as many books as the cart would neatly hold, a new scheme come into my head: which helped me over the two years' stile.

Without being of an awaricious temper, I like to be the owner of things. I shouldn't wish, for instance, to go partners with yourself in the Cheap Jack cart. It's not that I mistrust you, but that I'd rather know it was mine. Similarly, very likely you'd rather know it was yours. Well! A kind of jealousy began to creep into my mind when I reflected that all those books would have been read by other people long before they was read by her. It seemed to take away from her being the owner of 'em like. In this way the question got into my head:—Couldn't I have a book new-made express for her, which she should be the first to read?

It pleased me, that thought did, and * having formed the resolution, then come the question of a name. How did I hammer that hot iron into shape? This way. The most difficult explanation I had ever had with her, was, how I came to be called Doctor, and yet was no Doctor. We had first discovered the mistake we had dropped into, through her having asked me to prescribe for her when she had supposed me to be a Doctor in a medical point of view; so thinks I, 'Now, if I give this[1] book the name of my Prescriptions, and if she catches the idea that my only Prescriptions are for her amusement and interest—to make her laugh in a pleasant way—or to make her cry in a pleasant way—it will be a delightful proof to both of us that we have got over our difficulty.' It fell out to absolute perfection. †

But let me not anticipate. (I take that expression out of a lot of romances I bought for her. I never opened a single one of 'em—and I have opened many—but I found the romancer saying 'let me not anticipate.' Which being so, I wonder why he did anticipate, or who asked him to it.) Let me not, I say, anticipate. This same book took up all my spare time. At last it was done, and the two years' time was gone after all the other time before it, and where it's all gone to, Who knows? The new cart was finished—yellow outside, relieved with wermilion[2] and brass fittings

[1] '. . . this *miscellaneous* book . . .' (Wright).

[2] 'Now the words, "relieved with wermilion", *as* words, are not funny; and yet when Dickens is "relieved with wermilion" his face looks such unutterable things that even the most stoical fancies, as did Sophy herself once, that the Doctor is the *c-a-r-t*' (Field, p. 81).

—the old horse was put in it, a new 'un and a boy being laid on for the Cheap Jack cart—and I cleaned myself up to go and fetch her.

'Marigold,' says the gentleman, giving his hand hearty, 'I am very glad to see you.'

'Yet I have my doubts, sir,' says I,[1] 'if you can be half as glad to see me as I am to see you.'

'The time has appeared so long; has it, Marigold?'

'I won't say that, sir, considering its real length; but——'[2]

'What a start, my good fellow!'

Ah! I should think it was! Grown such a woman, so pretty, so intelligent, so expressive! I knew then that she must be really like my child, or I could never have known her, standing quiet by the door.

'[3]You are affected,' says the gentleman in a kindly manner.

'[4]I feel, sir,' says I, 'that I am but a rough chap in a sleeved waistcoat.'

'*I*[5] feel,' says the gentleman, 'that it was you who raised her from misery and degradation, and brought her into communication with her kind. But why do we converse alone together, when we can converse so well with her? Address her in your own way.'

'I am such a rough chap in a sleeved waistcoat, sir,' says I, 'and she is such a graceful woman, and she stands so quiet at the door!'

'Try if she moves at the old sign,' says the gentleman.

They had got it up together o' purpose to please me. For when I give her the old sign,[6] she rushed to my feet, and dropped upon her knees, holding up her hands to me with pouring tears of love and joy; and when I took her hands and lifted her, she clasped me round the neck and lay there; and I don't know what a fool I didn't make of myself, until we all three settled down into talking without sound, as if there was a something soft and pleasant spread over the whole world for us.[7]

CHAPTER II

EVERY item of my plan was crowned with success. Our reunited life was more than all that we had looked forward to. Content and joy went with us as the wheels of the two carts went round, and the same stopped

[1] 'Shaking hands' (Wright).
[2] 'Gives a start' (Wright).
[3] '*My poor fellow*, you are affected' (Wright).
[4] '*I don't know, sir*. I feel, sir . . .'; Marigold meanwhile 'Rubbing eyes and forehead and speaking pantingly and tremblingly' (Wright).
[5] *I* italic in *AYR* and *Berg*.
[6] 'Slaps table'; paragraph 'Feelingly and exultingly' (Wright).
[7] This ends ch. 1 of *Doctor Marigold's Prescriptions*, apart from a paragraph introducing the stories (mostly by other hands than Dickens's) which constituted the 'book' for Sophy. The Reading text resumes half-way through the first paragraph of ch. 8.

with us when the two carts stopped. I was as pleased and as proud as a Pug-Dog with his muzzle black-leaded for an evening party and his tail extra curled by machinery.

But I had left something out of my calculations. Now, what had I left out? To help you to a guess, I'll say, a figure. Come. Make a guess, and guess right. Nought? No. Nine? No. Eight? No. Seven? No. Six? No. Five? No. Four? No. Three? No. Two? No. One? No.[1] Now I'll tell you what I'll do with you. I'll say it's another sort of a figure altogether. There. Why then, says you, it's a mortal figure. No, nor yet a mortal figure. By such means you get yourself penned into a corner, and you can't help guessing a *im*mortal[2] figure. That's about it. Why didn't you say so sooner?

Yes. It was a immortal figure that I had altogether left out of my calculations. Neither man's nor woman's, but a child's. Girl's, or boy's? Boy's. 'I says the sparrow, with my bow and arrow.'[3] Now you have got it.

We were down at Lancaster, and I had done two nights' more than fair average business in the open square there. Mim's travelling giant otherwise Pickleson happened at the selfsame time to be a trying it on in the town. The genteel lay was adopted with him. No hint of a van. Green baize alcove leading up to Pickleson in a Auction Room. Printed poster, 'Free list suspended, with the exception of that proud boast of an enlightened country, a free press. Schools admitted by private arrangement. Nothing to raise a blush in the cheek of youth or shock the most fastidious.' Mim swearing most horrible and terrific[4] in a pink calico pay-place, at the slackness of the public. Serious hand-bill in the shops, importing that it was all but impossible to come to a right understanding of the history of David without seeing Pickleson.

I went to the Auction Room in question, and I found it entirely empty of everything but echoes and mouldiness, with the single exception of Pickleson on a piece of red drugget. This suited my purpose, as I wanted a private and confidential word with him: which was: 'Pickleson. Owing much happiness to you, I did put you in my will for a fypunnote; but, to save trouble here's four-punten down, which may equally suit your views, and let us so conclude the transaction.' Pickleson, who up to that remark had had the dejected appearance of a long Roman rushlight that couldn't anyhow get lighted, brightened up at his top extremity and made his acknowledgments in a way which (for him) was parliamentary eloquence. †

But what was to the present point in the remarks of the travelling

[1] 'Shaking head' (Wright).
[2] *im* italic in *AYR* and *Berg*.
[3] 'As if shooting an arrow' (Wright).
[4] Wright underlines doubly *horrible* and *terrific*.

giant otherwise Pickleson, was this: 'Doctor Marigold'—I give his words without a hope of conweying their feebleness—'who is the strange young man that hangs about your carts?'[1]—'The strange young *man?*'[2] I gives him back, thinking he meant her, and his languid circulation had dropped a syllable. 'Doctor,' he returns, with a pathos calculated to draw a tear from even a manly eye, 'I am weak, but not so weak yet as that I don't know my words. I repeat them, Doctor. The strange young man.' It then appeared that Pickleson being forced to stretch his legs (not that they wanted it) only at times when he couldn't be seen for nothing, to wit in the dead of the night and towards daybreak, had twice seen hanging about my carts, in that same town of Lancaster where I had been only two nights, this same unknown young man.

It put me rather out of sorts. Howsoever, I made light of it to Pickleson, and I took leave of Pickleson. Towards morning I kept a look-out for the strange young man, and what was more—I saw the strange young man. He was well-dressed and well-looking. He loitered very nigh my carts, watching them like, as if he was taking care of them, and soon after daybreak turned and went away. I sent a hail after him,[3] but he never started nor looked round, nor took the smallest notice.

We left Lancaster within an hour or two, on our way towards Carlisle. Next morning at daybreak I looked out again for the strange young man. I did not see him. But next morning I looked out again and there he was once more. I sent another hail after him, but as before he gave not the slightest sign of being anyways disturbed. This put a thought into my head. Acting on it, I watched him in different manners and at different times not necessary to enter into, till I found that this strange young man was deaf and dumb.

The discovery turned me over, because I knew that a part of that establishment where she had been, was allotted to young men (some of them well off), and I thought to myself 'If she favours him, where am I, and where is all that I have worked and planned for?' Hoping—I must confess to the selfishness—that she might *not*[4] favour him, I set myself to find out. At last I was by accident present at a meeting between them in the open air, looking on, leaning behind a fir-tree without their knowing of it. It was a moving meeting for all the three parties concerned. I knew every syllable that passed between them as well as they did.

[1] Kate Field quotes Pickleson's words, 'without a hope of conveying their feebleness. . . .Dickens outdoes himself. The contrast between the giant's purple face, swelling with effort, and the trickle of sound squeezed out at the risk of breaking every blood-vessel in Pickleson's head, is absolute perfection. A mountain never brought forth a smaller mouse, nor one that was so much worth the trouble' (p. 82). 'Giant's voice very high and very feeble' (Wright).

[2] *man* italic in *AYR* and *Berg*. 'Start and speak crossly, surprised' (Wright).

[3] 'I sent a "Aye" loud and high after him' (Wright).

[4] *not* italic in *AYR* and *Berg*.

I listened with my eyes, which had come to be as quick and true with deaf and dumb conversation as my ears with the talk of people that can speak.[1] He was going out to China as clerk in a merchant's house, which his father had been before him. He was in circumstances to keep a wife, and he wanted her to marry him and go along with him. She persisted—no. He asked if she didn't love him? Yes, she loved him dearly, dearly, but she could never disappoint her beloved good noble generous and I don't-know-what-all father (meaning me, the Cheap Jack in the sleeved waistcoat), and she would stay with him, Heaven bless him, though it was to break her heart! Then she cried most bitterly, and that made up my mind.

While my mind had been in an unsettled state about her favouring this young man, I had felt that unreasonable towards Pickleson, that it was well for him he had got his legacy down. For I often thought 'If it hadn't been for this same weak-minded giant, I might never have come to trouble my head and wex my soul about the young man.' But, once that I knew she loved him—once that I had seen her weep for him—it was a different thing. I made it right in my mind with Pickleson on the spot, and I shook myself together to do what was right by all.

She had left the young man by that time (for it took a few minutes to get me thoroughly well shook together), and the young man was leaning against another of the fir-trees—of which there was a cluster—with his face upon his arm. I touched him on the back. Looking up and seeing me, he says, in our deaf and dumb talk: 'Do not be angry.'

'I am not angry, good boy. I am your friend. Come with me.'

I left him at the foot of the steps of the Library Cart, and I went up alone. She was drying her eyes.

'You have been crying, my dear.'

'Yes, father.'

'Why?'

'A head-ache.'

'Not a heart-ache?'

'I said a head-ache, father.'

'Doctor Marigold must prescribe for that head-ache.'

She took up the book of my Prescriptions,[2] and held it up with a forced smile; but seeing me keep still and look earnest, she softly laid it down again, and her eyes were very attentive.

'The Prescription is not there, Sophy.'

'Where is it?'

'Here, my dear.'

I brought her young husband in, and I put her hand in his, and

[1] 'Low, monotonously' (Wright).
[2] 'Taking book up' (Wright).

my only further words to both of them were these: 'Doctor Marigold's last prescription. To be taken for life.'[1] After which I bolted.

When the wedding come off, I mounted a coat (blue, and bright buttons), for the first and last time in all my days, and I give Sophy away with my own hand. There were only us three and the gentleman who had had charge of her for those two years. I give the wedding dinner of four in the Library Cart. Pigeon pie, a leg of pickled pork, a pair of fowls, and suitable garden-stuff. The best of drinks. I give them a speech, and the gentlemen give us a speech, and all our jokes told, and the whole thing went off like a sky-rocket. In the course of the entertainment I explained to Sophy that I should keep the Library Cart as my living-cart when not upon the road, and that I should keep all her books for her just as they stood, till she come back to claim them. So she went to China with her young husband, and it was a parting sorrowful and heavy, and I got the boy I had another service, and so as of old, when my child and wife were gone, I went plodding along alone, with my whip over my shoulder, at the old horse's head.

Sophy wrote me many letters, and I wrote her many letters. About the end of the first year she sent me one in an unsteady hand:[2] 'Dearest father, not a week ago I had a darling little daughter, but I am so well that they let me write these words to you. Dearest and best father, I hope my child may not be deaf and dumb, but I do not yet know.' When I wrote back, I hinted the question; but as Sophy never answered that question, I felt it to be a sad one, and I never repeated it. For a long time our letters were regular, but then they got irregular through Sophy's husband being moved to another station and through my being always on the move. But we were in one another's thoughts, I was equally sure, letters or no letters.

Five years, odd months, had gone since Sophy went away. I was at a greater heighth of popularity than ever. I had had a first-rate autumn of it, and on the twenty-third of December, one thousand eight hundred and sixty-four, I found myself at Uxbridge, Middlesex,[3] clean sold out. So I jogged up to London with the old horse, light and easy, to have my Christmas-Eve and Christmas-Day alone by the fire in the Library Cart, and then to buy a regular new stock of goods all round, to sell 'em again and get the money.

I am a neat hand at cookery, and I'll tell you what I knocked up for my Christmas-Eve dinner in the Library Cart. I knocked up a beefsteak pudding for one, with two kidneys, a dozen oysters, and a couple of mushrooms, thrown in. It's a pudding to put a man in good humour

[1] 'Slaps hand into left and shaking up and down and in [and] out'; at the end of paragraph, 'Wipe tear from eye' (Wright).

[2] 'Reading letter as if in left hand' (Wright).

[3] '20 miles from London' (Wright inserts).

with everything, except the two bottom buttons of his waistcoat. Having relished that pudding and cleared away, I turned the lamp low, and sat down by the light of the fire, watching it as it shone upon the backs of Sophy's books.

Sophy's books so brought up Sophy's self, that I saw her touching face quite plainly, before I dropped off dozing by the fire. This may be a reason why Sophy, with her deaf and dumb child in her arms, seemed to stand silent by me all through my nap. Even when I woke with a start, she seemed to vanish, as if she had stood by me in that very place only a single instant before.

I had started at a real sound, and the sound was on the steps of the cart. It was the light hurried tread of a child coming clambering up. That tread of a child had once been so familiar to me, that for half a moment I believed I was a going to see a little ghost.[1]

But the touch of a real child was laid upon the outer handle of the door, and the handle turned and the door opened a little way, and a real child peeped in. A bright little comely girl with large dark eyes.

Looking full at me, the tiny creature took off her mite of a straw hat, and a quantity of dark curls fell all about her face. Then she opened her lips, and said in a pretty voice:

'Grandfather!'[2]

'Ah my God!' I cries out. 'She can speak!'

In a moment Sophy was round my neck as well as the child, and her husband was a wringing my hand with his face hid, and we all had to shake ourselves together before we could get over it.[3] And when we did begin to get over it, and I saw the pretty child a talking, pleased and quick and eager and busy, to her mother, in the signs that I had first taught her mother, the happy and yet pitying tears fell rolling down my face.[4]

THE END

[1] 'Left hand to forehead [speak?] in a whisper'; at the beginning of the next paragraph, 'Dickens comes to Left end of table' (Wright).

[2] 'Child's voice' (Wright). 'And when the prattler had broken the still solitariness of the caravan with the call "Grandfather", the exclamation of the elated old man brought forth numerous pocket handkerchiefs which were still in use when genuine applause followed the talented reader into the ante-room' (*Staffordshire Sentinel*, 4 May 1867).

[3] 'Left Hand to forehead, head leaning to left. Right Hand thrust out and shaking young man's hand' (Wright).

[4] '. . . those tears steal into our eyes as well; and when Dickens steals away, there seems to be more love and unselfishness in the world than before we took Doctor Marigold's prescription' (Field, p. 83).

SIKES AND NANCY

'I HAVE been trying, alone by myself, the Oliver Twist murder', Dickens told a friend in 1863, 'but have got something so horrible out of it that I am afraid to try it in public' (*N*, iii. 353). Five years later, when preparing his Farewell tour, he revived the notion, partly because a new attraction would ensure that his impresarios would not lose on their liberal payments to him, but also because he wanted 'to leave behind me the recollection of something very passionate and dramatic, done with simple means, if the art would justify the theme' (*N*, iii. 679). He remained, however, so uncertain whether to 'try it in public' that he mounted an elaborate 'trial performance' before a hundred distinguished guests, whom he flabbergasted by the brilliance and intensity of his acting. All his guests were 'unmistakably pale, and had horror-struck faces', he noted gleefully, and he was similarly exultant when William Harness, the Shakespearean scholar, wrote saying that it was 'a most amazing and terrific thing' and added:

'but I am bound to tell you that I had an almost irresistible impulse upon me to *scream*, and that, if anyone had cried out, I am certain I should have followed.' He had no idea that, on the night, Priestley, the great ladies' doctor, had taken me aside and said: 'My dear Dickens, you may rely upon it that if one woman cries out when you murder the girl, there will be a contagion of hysteria all over this place.' It is impossible to soften it without spoiling it, and you may suppose that I am rather anxious to discover how it goes on the Fifth of January !!! . . . I asked Mrs. Keeley, the famous actress, who was at the experiment: 'What do *you* say? Do it or not?' 'Why, of course, do it,' she replied. 'Having got at such an effect as that, it must be done. But,' rolling her large black eyes very slowly, and speaking very distinctly, 'the public have been looking out for a sensation these last fifty years or so, and by Heaven they have got it!' With which words, and a long breath and a long stare, she became speechless. Again, you may suppose that I am a little anxious! (*N*, iii. 687)

Members of Dickens's family and entourage, however, urged him not to do it: they feared for its effect upon his health (rightly, as it proved), or they found such outright histrionic violence disquieting, distasteful or undignified. But, as his manager George Dolby had already found, and was often to find later, 'he would listen to no remonstrance in respect of it' (Dolby, p. 344), and when the item entered his repertoire, he became angry over suggestions that he should limit the number of performances. Dolby tells some extraordinary stories about Dickens's obsession with this item—tells also that, afterwards, Dickens confessed that it had been 'madness ever to have given the "Murder" Reading, under the conditions of a travelling life, and worse than madness to have given it with such frequency' (p. 442). Wilkie Collins was not alone among his friends in believing that this Reading—undertaken at a time when his health was clearly failing—did more to kill him than all his work put together.

Collins, however, had encouraged him to do the Reading, and in a longer form than at the trial performance. The narrative then had consisted of three chapters—I and II (mostly from chs. 45 and 46 of *Oliver Twist*) concerned Fagin's setting Bolter on to 'dodge' Nancy, and Bolter's eavesdropping on her meeting with Mr. Brownlow and Rose Maylie; in Chapter III (from ch. 47 and the opening page of ch. 48), Bill Sikes, incensed by Fagin, murders Nancy: dawn breaks and he leaves the house. Collins, with Charles Kent, urged Dickens to continue to the death of Sikes. After rejecting this as more than any audience would stand, Dickens complied, writing into the endpapers of his prompt-book (*Berg*) a brilliantly condensed version of chs. 48 and 50, with some unusually radical re-writing so that the action is seen through the eyes of the terrified Sikes. He had a new edition printed from this much-revised prompt-copy (a facsimile of this edition was edited by J. H. Stonehouse in 1921). His own prompt-copy of this new edition has been lost, but fortunately the actress Adeline Billington transcribed into a copy, which he had given her, Dickens's own elaborate underlinings, long dashes, stage-directions, etc.: and this Billington copy (*Suzannet*) is the text printed below. Some particulars of the *Berg* copy are given in the footnotes. A facsimile of *Suzannet*, with introduction by Philip Collins, has been published by Dickens House (1982).

'Never, probably, through the force of mere reading was a vast concourse of people held so completely within the grasp of one man', reported *The Times* (8 January 1869) on his first public performance. 'Every personage in the tale is played with a distinctness that belongs to the highest order of acting.' Most critics agreed that this was Dickens's most remarkable platform feat, a performance which 'our greatest histrionic artists might deem it the height of their ambition to produce' (*Daily Telegraph*, 6 January 1869)—and so indeed said the greatest of histrionic artists, Macready, who gasped out, to Dickens's gratified ears, ' "In my—er—best times—er—you remember them, my dear boy—er—gone, gone!—no," with great emphasis again,— "it comes to this—er—TWO MACBETHS!" with extraordinary energy' (*N*, iii. 704). It was 'a masterpiece of reading, quite unparalleled in its way', a Dublin paper said, praising particularly his impersonation of Nancy, and it proved that 'Mr. Dickens is the greatest reader of the greatest writer of the age'. He took on the physical shape, as well as the voice, of Fagin: 'He has always trembled on the boundary line that separates the reader from the actor; in this case he clears it by a leap' (*Times*, 17 November 1868). His young colleague Edmund Yates records that when

> . . . gradually warming with excitement, he flung aside his book and acted the scene of the murder, shrieked the terrified pleadings of the girl, growled the brutal savagery of the murderer, brought looks, tones, gestures simultaneously into play to illustrate his meaning, there was not one, not even of those who had

known him best or who believed in him most, but was astonished at the power and the versatility of his genius. (*Tinsley's Magazine*, iv (1869), 62)

'I shall tear myself to pieces', he whispered to Charles Kent as he made his way to the platform for his last performance of *Sikes and Nancy*, on 8 March 1870 (Kent, p. 87: and Kent's chapter on this Reading is of particular merit). He had already done so, too many times. During his final Farewells his physician was in constant attendance, and recorded that *Sikes* made his pulse-rate rise from 72 to 124. After reading it, he would often lie on a sofa, unable to speak a word, during a ten-minute interval, before staggering back on-stage to read a more cheerful item which would send his audience home happy. Three months after this Farewell season, he was dead: but a friend reports that, a day or two before his death, he was discovered in the grounds at Gad's Hill re-enacting the murder of Nancy.

Sikes and Nancy

CHAPTER I

Fagin the receiver of stolen goods was up, betimes, one morning, and waited impatiently for the appearance of his new associate, Noah Claypole, otherwise Morris Bolter; who at length presented himself, and, cutting a monstrous slice of bread, commenced a voracious assault on the breakfast.[1]

'Bolter, *Bolter*.'

'Well, here I am. What's the matter? Don't yer ask me to do anything till I have done eating. That's a great fault in this place. Yer never get time enough over yer meals.'

'You can talk as you eat, can't you?'

'Oh yes, I can talk. I get on better when I talk. *Talk away*. Yer won't interrupt me.'

There seemed, indeed, no great fear of anything interrupting him, as he had evidently sat down with a determination to do a deal of business. *

'I want you, Bolter,' *leaning over the table*, 'to do a piece of work for me, my dear, that needs great care and caution.'

'I say, don't yer go a-shoving me into danger, yer know. That don't suit me, that don't; and so I tell yer.'

'There's not the smallest danger in it—not the very smallest; it's only to *dodge a woman*.'[2]

'An old woman?'[3]

'A young one.'

'I can do that pretty well. I was a regular sneak when I was at school. What am I to dodge her for? Not to——'

'Not to do anything, but tell me where she goes, who she sees, and, if possible, what she says; to remember the street, if it is a street, or the house, if it is a house; and to bring me back all the information you can.'

'What'll yer give me?'

'If you do it well, a pound, my dear. One pound. And that's what I never gave yet, for any job of work where there wasn't valuable consideration to be got.'

'Who is she?'

[1] On the character of 'the half-knowing, half-stupid Claypole, . . . a light is thrown by the "reading" that is scarcely to be found in the book'. This opening scene 'caused now and then a little mirth' (*The Times*, 8 January 1869).

[2] *woman* doubly underlined.

[3] Question-mark (as in the novel and in *Berg*) editorially supplied.

'One of us.'

'Oh Lor! Yer doubtful of her, are yer?'

'She has found out some new friends, my dear, and I must know who they are.'

'I see. Ha! ha! ha! I'm your man. Where is she? Where am I to wait for her? Where am I to go?'

'All that, my dear, you shall hear from me. I'll point her out at the proper time. You keep ready, in the clothes I have got here for you, and leave the rest to me.'

That night, and the next, and the next again, the spy sat booted and equipped in the disguise of a carter: ready to turn out at a word from Fagin. Six nights passed, and on each, Fagin came home with a disappointed face, and briefly intimated that it was not yet time. On the seventh he returned exultant. It was Sunday Night.

'She goes abroad to-night,' said Fagin, 'and on the right errand, I'm sure; for she has been alone all day, and the man she is afraid of will not be back much before daybreak. Come with me! Quick!'

They left the house, and, stealing through a labyrinth of streets, arrived at length before a public-house. It was past eleven o'clock, and the door was closed; but it opened softly as Fagin gave a low whistle. They entered, without noise.

Scarcely venturing to whisper, but substituting dumb show for words, Fagin pointed out a pane of glass high in the wall to Noah, and signed to him to climb up, on a piece of furniture below it, and observe the person in the adjoining room.

'Is that the woman?'

Fagin nodded 'yes'.

'I can't see her face well. She is looking down, and the candle is behind her.'

'Stay there.' He signed to the lad, who had opened the house-door to them; who withdrew—entered the room adjoining, and, under pretence of snuffing the candle, moved it in the required position; then he spoke to the girl, causing her to raise her face.

'I see her now!'

'Plainly?'

'I should know her among a thousand.'[1]

The spy descended, the room-door opened, and the girl came out. Fagin drew him behind a small partition, and they held their breath as she passed within a few feet of their place of concealment, and emerged by the door at which they had entered.

'*After her*!! To the *left*. Take the left hand, and keep on the other side. *After her*!!'[2]

[1] Marginal stage-direction *Beckon down.*

[2] All underlinings double in this paragraph: also for *She looked nervously round*, below.

The spy darted off; and, by the light of the street lamps, saw the girl's retreating figure, already at some distance before him. He advanced as near as he considered prudent, and kept on the opposite side of the street. *She looked nervously round.* She seemed to gather courage as she advanced, and to walk with a steadier and firmer step. The spy preserved the same relative distance between them, and followed.

CHAPTER II

THE churches chimed three quarters past eleven, as the two figures emerged on London Bridge. The young woman advanced with a swift and rapid step, and looked about her as though in quest of some expected object; the young man, who slunk along in the deepest shadow he could find, and, at some distance, accommodated his pace to hers: stopping when she stopped: and as she moved again, creeping stealthily on: but never allowing himself, in the ardour of his pursuit, to gain upon her. Thus, they crossed the bridge, from the Middlesex to the Surrey shore, when the woman, disappointed in her anxious scrutiny of the foot-passengers, turned back. The movement was sudden; but the man was not thrown off his guard by it; for, shrinking into one of the recesses which surmount the piers of the bridge, and leaning over the parapet the better to conceal his figure, he suffered her to pass. When she was about the same distance in advance as she had been before, he slipped quietly down, and followed her again. At nearly the centre of the bridge she stopped. He stopped.

It was a very dark night. The day had been unfavourable, and at that hour and place there were few people stirring. Such as there were, hurried past: possibly without seeing, certainly without noticing, either the woman, or the man. Their appearance was not attractive of such of London's destitute population, as chanced to take their way over the bridge that night; and they stood there in silence: neither speaking nor spoken to.[1]

The girl had taken a few turns to and fro—closely watched by her hidden observer—*when the heavy bell of St. Paul's tolled for the death of another day. Midnight had come upon the crowded city. Upon the palace, the night-cellar, the jail, the madhouse: the chambers of birth and death, of health and sickness, upon the rigid face of the corpse and the calm sleep of the child.*

A young lady, accompanied by a grey-haired gentleman, alighted from a hackney-carriage. They had scarcely set foot upon the pavement of the bridge, when the girl started, and joined them.

'*Not here*!! I am afraid to speak to you here. Come away—out of the public road—down the steps yonder!'[2]

[1] *Berg* here included the next paragraph in the novel ('A mist hung over the river . . .'), until Dickens deleted it. [2] Marginal stage-direction *Point R*[*ight*].

The steps to which she pointed, were those which, on the Surrey bank, and on the same side of the bridge as Saint Saviour's Church, form a landing-stairs from the river. To this spot the spy hastened unobserved; and after a moment's survey of the place, he began to descend.

These stairs are a part of the bridge; they consist of three flights. Just below the end of the second, going down, the stone wall on the left terminates in an ornamental pilaster facing towards the Thames.[1] At this point the lower steps widen: so that a person turning that angle of the wall, is necessarily unseen by any others on the stairs who chance to be above, if only a step. The spy looked hastily round, when he reached this point; and as there seemed no better place of concealment, and as the tide being out there was plenty of room, he slipped aside, with his back to the pilaster, and there waited: pretty certain that they would come no lower down.

So tardily went the time in this lonely place, and so eager was the spy, that he was on the point of emerging from his hiding-place, and regaining the road above, when he heard the sound of footsteps, and directly afterwards of voices almost close at his ear.

He drew himself straight upright against the wall, and listened attentively.

'This is far enough,' *said a voice, which was evidently that of the gentleman.* 'I will not suffer the young lady to go any further. Many people would have distrusted you too much to have come even so far, but you see I am willing to humour you.'

'To humour me!' *cried the voice of the girl* whom he had followed.[2] 'You're considerate, indeed, sir. To humour me! Well, well, it's no matter.'

'Why, for what purpose can you have brought us to this strange place? Why not have let me speak to you, above there, where it is light, and there is something stirring, instead of bringing us to this dark and dismal hole?'

'I told you before, that I was afraid to speak to you there. I don't know why it is,' *said the girl shuddering,*[3] 'but I have such a fear and dread upon me to-night that I can hardly stand.'

'A fear of what?'

'I scarcely know of what—I wish I did. Horrible thoughts of *death*—and *shrouds* with *blood*[4] upon them—and a fear that has made me burn as if I was on fire—have been upon me all day. I was reading a book to-night, to while the time away, and the same things came into the print.'

[1] Sir Frederick Pollock noted how graphically Dickens conjured up this London Bridge scene by gestures and voice: 'What an actor he would have made! what a success he must have had if he had gone to the bar!' (*Personal Remembrances* (1887), ii. 199).

[2] The underlining should probably have continued to the end of this sentence, as in *Berg*.

[3] Marginal stage-direction *Shudder*.

[4] *blood* doubly underlined.

'Imagination!'

'No imagination. I swear I saw "*coffin*"[1] written in every page of the book in large black letters,—aye, and they carried one close to me, in the streets to-night.'

'There is nothing unusual in that. They have passed me often.'

'*Real ones*.[2] This was not.'

'Pray speak to her kindly,' said the young lady to the grey-haired gentleman. 'Poor creature! She seems to need it.'

'Bless you, miss, for that! Your haughty religious people would have held their heads up to see me as I am to-night, and would have preached of flames and vengeance. Oh,[3] dear lady, why ar'n't those who claim to be God's own folks, as gentle and as kind to us poor wretches as you!'[4]

—'You were not here last Sunday night, girl, as you appointed.'

'I couldn't come. I was kept by force.'

'By whom?'

'*Bill*——*Sikes*[5]—him that I told the young lady of before.'

'You were not suspected of holding any communication with anybody on the subject which has brought us here to-night, I hope?'

'No,' replied the girl, shaking her head.[6] 'It's not very easy for me to leave him unless he knows why; I couldn't have seen the lady when I did, but that I gave him a drink of *laudanum* before I came away.'

'Did he awake before you returned?'

'No; and neither he nor any of them suspect me.'

'Good. Now listen to me. I am Mr. Brownlow, this young lady's friend. I wish you, in this young lady's interest, and for her sake, to deliver up Fagin.'

'Fagin! I will not do it! I will never do it! Devil that he is, and worse than devil as he has been to me, as my Teacher in all Devilry, I will never do it.'

'Why?'

'For the reason that, bad life as he has led, I have led a bad life too; for the reason that there are many of us who have kept the same courses together, and I'll not turn upon them, who might—any of them—have turned upon me, but didn't, bad as they are. Last, for the reason—(*how*

[1] *coffin* doubly underlined, and enclosed in double quotation-marks.

[2] Italic in the novel, and in *Berg* and *Suzannet*, where it is also doubly underlined.

[3] *Oh* in the novel, altered to *Ah* in *Berg*.

[4] In *Berg*, Mr. Brownlow's speech here about 'the Mussulman' was printed, amended, and finally deleted.

[5] Double dash after *Bill*, and *Sikes* is doubly underlined.

[6] In *Berg*, the phrase 'replied the girl, shaking her head' is underlined. There are other such occasions where underlining is used in *Berg*, more often than in *Suzannet*, simply to isolate a narrative phrase inside a dialogue; they will not be mentioned hereafter.

can I say it with the young lady here!)[1]—that, among them, there is one—*this Bill—this Sikes*—the most desperate of all—*that I can't leave.*[2] * Whether it is God's wrath for the wrong I have done, I don't know, but I am drawn back to him through everything, and I should be, I believe, if I knew that I was to *die* by his hand!'

'But, put one man—not him—not one of the gang—the one man Monks into my hands, and leave him to me to deal with.'

'What if he turns against the others?'

'I promise you that, in that case, there the matter shall rest; they shall go scot free.'

'Have I the lady's promise for that?'

'You have,' replied Rose Maylie, the young lady.

'I have been a liar, and among liars from a little child, but I will take your words.'

After receiving an assurance from both, that she might safely do so, she proceeded in a voice so low that it was often difficult for the listener to discover even the purport of what she said, to describe the means by which this one man Monks might be found and taken. * But nothing would have induced her to compromise one of her own companions; little reason though she had, poor wretch! to spare them.

'Now,' said the gentleman, when she had finished, 'you have given us most valuable assistance, young woman, and I wish you to be the better for it. What can I do to serve you?'

'Nothing.'

'You will not persist in saying that; think now; take time. Tell me.'

'Nothing, sir. You can do nothing to help me. I am past all hope.'

'You put yourself beyond the pale of hope. The past has been a dreary waste with you, of youthful energies mis-spent, and such treasures lavished, as the Creator bestows but once and never grants again, but, *for the future, you may hope*![3] [I do not say that it is in our power to offer you peace of heart and mind, for that must come as you seek it; but a quiet asylum, either in England, or, if you fear to remain here, in some foreign country, it is not only within the compass of our ability but our most anxious wish to secure you. Before the dawn of morning, before this river wakes to the first glimpse of daylight, you shall be placed as entirely beyond the reach of your former associates, and leave as complete an absence of all trace behind you, as if you were to disappear from the earth this moment.] Come! I would not have you go back to exchange one word

[1] Double vertical lines in the margin against the bracketed interjection. The following sentences are taken from ch. 40.

[2] *that I can't leave*, and *die* below, doubly underlined.

[3] *future* doubly underlined. A pencilled footnote (not, it would seem, in the handwriting of either Adeline Billington or John Hollingshead) reads: 'The passage in square brackets deleted'. The passage is deleted, in the same ink as the rest of the manuscript underlinings and words.

with any old companion, or take one look at any old haunt. Quit them all, while there is time and opportunity!'

'She will be persuaded now,' cried the young lady.

'I fear not, my dear.'

'*No, sir—no, miss.*[1] I am chained to my old life. I *loathe* and *hate* it, but I cannot *leave* it.—When ladies as young and good, as happy and beautiful as you, miss, give away your hearts, love will carry even you all lengths.[2] When such as I, who have no certain roof but the coffin-lid, and no friend in sickness or death but the hospital-nurse, set our rotten hearts on any man, who can hope to cure us![3]—This fear comes over me again. I must go home. Let us part. I shall be watched or seen.[4] *Go! Go!* If I have done you any service, all I ask is, leave me, and let me go my way alone.'

'Take this purse,' cried the young lady.[5] 'Take it for my sake, that you may have some resource in an hour of need and trouble.'

'*No!* I have not done this for *money. Let me have that to think of.*[6] And yet——give me something that you have worn—I should like to have something—*no, no,* not a *ring,* they'd rob me of that—your *gloves* or *handkerchief*—anything that I can keep, as having belonged to you. There. *Bless you! God bless you!! Good-night, good-night!*'

The agitation of the girl, and the apprehension of some discovery which would subject her to violence, seemed to determine the gentleman to leave her. The sound of retreating footsteps followed, and the voices ceased. *

After a time Nancy ascended to the street.[7] The spy remained on his post for some minutes, and then, *after peeping out,* to make sure that he was unobserved, darted away, and made for Fagin's house as fast as his legs would carry him.

[1] *No, sir* and *miss* doubly underlined. The novel does not contain 'no, miss'. This insertion, like other differences between the novel and the Reading texts, enlarges Rose Maylie's prominence in this encounter.

[2] In *Berg* a marginal stage-direction is written against the next sentence—*Start coming.* The 'Start' presumably came at 'This fear comes over me . . .'

[3] In *Berg*, the next sentence from ch. 40 is here inserted in Dickens's handwriting, and subsequently deleted: 'Pity us, Lady! pity us, for having only one feeling of the woman left, and for having that turned—by a heavy judgment—from a comfort and a pride, into a new means of suffering.—'

[4] Marginal stage-direction *Look Round with Terror.* The following *Go! Go!* is doubly underlined.

[5] Marginal stage-direction *Action.*

[6] In the remainder of this paragraph, there is a double dash after 'And yet'; *ring* is trebly underlined; *handkerchief* and *God bless you!! Good-night, good-night!* are doubly underlined. The dash after 'they'd rob me of that' (a manuscript insertion) is editorially supplied.

[7] In the margin there is a vertical line against the remainder of this paragraph.

CHAPTER III

IT was nearly two hours before daybreak; that time which in the autumn of the year, may be truly called the dead of night; when the streets are silent and deserted; when even sound appears to slumber, and profligacy and riot have staggered home to dream; it was at this still and silent hour, that Fagin sat in his old lair. Stretched upon a mattress on the floor, lay Noah Claypole, otherwise *Morris Bolter*, fast asleep. Towards him the old man sometimes directed his eyes for an instant, and then brought them back again to the wasting candle. *

He sat without changing his attitude, or appearing to take the smallest heed of time, until the door-bell rang. He crept up-stairs, and presently returned accompanied by a man muffled to the chin, who carried a bundle under one arm. Throwing back his outer coat, the man displayed the *burly frame of Sikes, the housebreaker.*[1]

'*There*!' laying the bundle on the table. 'Take care of that, and do the most you can with it. It's been trouble enough to get. I thought I should have been here three hours ago.'

Fagin laid his hand upon the bundle, and locked it in the cupboard.[2] But he did not take *his eyes off the robber, for an instant.*

'Wot now?' cried Sikes. 'Wot do you look at a man, like that, for?'

Fagin raised his right hand,[3] and shook his trembling forefinger in the air.

'Hallo!' *feeling in his breast.* 'He's gone mad. I must look to myself here.'

'No, no, it's not—you're not the person, Bill. I've no—no fault to find with you.'

'Oh! you haven't, haven't you?' *passing a pistol into a more convenient pocket.*[4] 'That's lucky—for one of us. Which one that is, don't matter.'

'I've got that to tell you, Bill, will make you worse than me.'

'Aye? Tell away! Look sharp, or Nance will think I'm lost.'

'*Lost*! She has pretty well settled that, in her own mind, already.'

He looked, perplexed, into the old man's face, and reading no satisfactory explanation of the riddle there, clenched his coat collar in his huge hand and shook him soundly.

'Speak, will you? Or if you don't, it shall be for want of breath. Open

[1] *Sikes* doubly underlined.

[2] Marginal stage-direction *Cupboard action.* Just below, *an instant* is doubly underlined.

[3] Marginal stage-direction *Action.* In *Berg*, this paragraph has double vertical lines against it in the margin. According to Kent (p. 261), 'Not a word of it was said. It was simply *done.*' And similarly about the next *Action*: 'Not a word was said about the pistol —the marginal direction was simply attended to.'

[4] Marginal stage-direction *Action.*

your mouth and say wot you've got to say. Out with it, you *thundering, blundering, wondering* old *cur*,[1] out with it!'

[2]'Suppose that lad that's lying there——' Fagin began.

Sikes turned round to where Noah was sleeping, as if he had not previously observed him. 'Well?'

'Suppose that lad was to peach—to blow upon us all. Suppose that lad was to do it, of his own fancy—not grabbed, tried, earwigged by the parson and brought to it on bread and water,—but of his own fancy; to please his own taste; stealing out at nights to do it. Do you hear me? Suppose he did all this, what then?'

'What then? If he was left alive till I came, I'd grind his skull under the iron heel of my boot into as many grains as there are hairs upon his head.'

'What if *I*[3] did it! *I*, that know so much, and could hang so many besides myself!'

'I don't know. I'd do something in the jail that 'ud get me put in irons; and, if I was tried along with you, I'd fall upon you with them in the open court, and beat your brains out afore the people. I'd smash your head as if a loaded waggon had gone over it.'

Fagin looked hard at the robber; and, motioning him to be silent,[4] stooped over the bed upon the floor, and shook the sleeper to rouse him.

'Bolter! Bolter! *Poor lad*!'[5] said Fagin, looking up with an expression of devilish anticipation, and speaking slowly and with marked emphasis. '*He's tired*—tired with watching for *her*[6] so long—watching for *her*, Bill.'

'Wot d'ye mean?'

Fagin made no answer, but bending over the sleeper again, hauled him into a sitting posture. When his assumed name had been repeated several times, Noah rubbed his eyes, and, giving a heavy yawn, looked sleepily about him.

'Tell me that again—once again, just for him to hear,' said the Jew, *pointing to Sikes* as he spoke.[7]

'Tell yer what?' *asked the sleepy Noah, shaking himself pettishly*.

'That about——NANCY!![8] You followed her?'

'Yes.'

[1] Underlining in crescendo: once for *thundering*, twice for *blundering*, thrice for *wondering*.

[2] Marginal stage-direction (*Points to Bed*). In *Berg*, this is *Bed Action*.

[3] Like the *I* in the next sentence, this is italic in the novel and in both Readings texts.

[4] Marginal stage-direction *Bed Stooping Action*.

[5] Doubly underlined.

[6] *her* . . . *her* italic in the novel and in the Readings texts; also, in both cases, doubly underlined.

[7] Marginal stage-direction *Pointing Action*. In *Berg*, which here has *Point Action*, there is a rough sketch of a hand with pointing forefinger.

[8] NANCY is in capitals thus in the novel and the Readings texts; in *Suzannet* it is doubly underlined.

'To London Bridge?'

'Yes.'

'Where she met two people?'

'So she did.'

'A gentleman and a lady that she had gone to of her own accord before, who asked her to give up all her pals, and Monks first, which *she did*—and to describe him, which *she did*—and to tell her what house it was that we meet at, and go to, which *she did*—and where it could be best watched from, which *she did*—and what time the people went there, which *she did*. *She did all this.*[1] She told it *all*, every word, without a threat, without a murmur—*she did—did she not*?'

'All right,' *replied Noah, scratching his head.*[2] 'That's just what it was!'

'What did they say about last Sunday?'

'About last Sunday! Why, I told yer that before.'

'Again. *Tell it again*!'

'They asked her,' as he grew more wakeful, and seemed to have a dawning perception who Sikes was, 'they asked her why she didn't come, last Sunday, as she promised. She said she couldn't.'

'*Why*? Tell *him that*.'[3]

'Because she was forcibly kept at home by Bill—Sikes——the man that she had told them of before.'

'What more of him? What more of Bill—Sikes—the man she had told them of before? Tell him that, *tell him that*.'

'Why, that she couldn't very easily get out of doors unless he knew where she was going to, and so the first time she went to see the lady, she —*ha*! *ha*! *ha*![4] it made me *laugh* when she said it, *that* did—she gave him, a drink *of laudanum*!! ha! ha! ha!'[5]

Sikes rushed from the room, and darted up the stairs.

'Bill, *Bill*!'[6] cried Fagin, following him, hastily. 'A word. Only a word.'

'Let me out. Don't *speak* to me! it's not *safe*. *Let me out*.'

'Hear me speak a word,' rejoined Fagin, *laying his hand upon the lock.*[7] 'You won't be——you won't be——*too*—*violent*, Bill?'

The day was breaking, and there was light enough for the men to see

[1] *She did all this* doubly underlined; so, below, are *all* and *she did—did*.

[2] Marginal stage-direction *Sleepy Action*.

[3] *Why* and *him* doubly underlined. After 'Tell *him that*' there is a manuscript insertion, deleted, of a repeated 'Tell *him* that'.

[4] Doubly underlined; so, below, are *that* and *laudanum*.

[5] Dickens wrote to Georgina Hogarth on 7 March 1869: 'As always happens now—and did not at first—they [the Manchester audience] were unanimously taken by Noah Claypole's laugh' (*N*, iii. 710). The concluding 'ha! ha! ha!' is added in handwriting in *Suzannet*.

[6] Doubly underlined.

[7] Marginal stage-direction *Action*.

each other's faces. They exchanged a brief glance; there was the same fire in the eyes of both.[1]

'I mean, not too——*violent*——for——for——*safety*. Be *crafty*, Bill, and not too *bold*.'[2]

The robber dashed into the silent streets.

Without one pause, or moment's consideration; without once turning his head to the right or left; without once raising his eyes to the sky, or lowering them to the ground, but looking straight before him with savage resolution: he muttered not a word, nor relaxed a muscle, until he reached his own house-door.——He opened it, *softly*,[3] with a key; strode lightly up the stairs; and entering his own room, *double-locked the door, and drew back the curtain of the bed.*[4]

The girl was lying, half-dressed, upon the bed. He had roused her from her sleep, for she raised herself with a hurried and startled look.

'Get up!'

'It *is*[5] you, Bill!'

'*Get up*!!!'[6]

There was a candle burning, but he drew it from the candlestick, and hurled it under the grate. Seeing the faint light of early day without, the girl rose to undraw the curtain.

'*Let it be*. There's light enough for wot I've got to do.'——[7]

'*Bill, why do you look like that at me?*'[8]

The robber regarded her, for a few seconds, with dilated nostrils[9] *and heaving breast; then, grasping her by the head and throat, dragged her into the middle of the room, and placed his heavy hand upon her mouth.*

'You were watched to-night, *you she-devil; every word you said was heard*.'[1]

'Then if every word I said was heard, it was heard that I spared you. Bill, *dear Bill*, you cannot have the heart to kill me. Oh! think of all I have given up, only this one night, for *you*. Bill, *Bill*![2] For dear God's sake, for your own, for mine, stop before you *spill my blood*!!! I have

[1] In *Berg*, the whole paragraph is underlined, and 'there was the same fire in the eyes of both' doubly underlined.

[2] *safety* and *bold* doubly underlined.

[3] *softly* doubly underlined. *Suzannet* has here the marginal stage-direction *Murder coming*; in *Berg* it occurs, more plausibly, at the end of this paragraph.

[4] Marginal stage-direction *Action*.

[5] *is* italic in the novel and in both Readings texts; doubly underlined in *Suzannet*.

[6] Trebly underlined.

[7] *Let it be* doubly underlined. After '. . . got to do', a very long dash, followed by (*Pause*), trebly underlined.

[8] Whole speech has double interrupted underlining.

[9] *dilated nostrils* doubly underlined.

[1] *devil* trebly underlined; *she-* and *word* and *heard* doubly underlined.

[2] *Bill* and, below, *upon my guilty soul I have*, doubly underlined.

been *true* to you, *upon my guilty soul I have*!!! The gentleman and that dear lady told me to-night of a home in some foreign country where I could end my days in solitude and peace.[1] Let me see them again, and beg them, on my knees, to show the same mercy to you; and let us both leave this dreadful place, and far apart lead better lives, and forget how we have lived, except in prayers, and never see each other more. It is never too late to repent. They told me so—I feel it now. But we must have *time*—we must have a *little*, *little time*!'[2]

The housebreaker freed one arm, and grasped his pistol. The certainty of immediate detection if he fired, flashed across his mind; and he beat it *twice* upon the upturned face *that almost touched his own.*[3]

She staggered and fell, but raising herself on her knees, *she drew from her bosom a white handkerchief—Rose Maylie's*[4]*—and holding it up towards Heaven, breathed one prayer, for mercy to her Maker.*

It was a ghastly figure to look upon. The murderer staggering backward to the wall, and shutting out the sight with his hand, seized a heavy club, and *struck her* down!![5]

The bright sun burst upon the crowded city in clear and radiant glory. *Through costly-coloured glass and paper-mended window, through cathedral dome and rotten crevice*, it shed *its equal ray*. It lighted up *the room* where *the murdered woman* lay.[6] It did. He tried to shut it out, but *it would stream in*. If the sight had been a *ghastly* one in the *dull morning*, what was it, *now*, in all that *brilliant light*!!![7]

He had not moved; he had been afraid to stir. There had been a moan and motion of the hand; and, with terror added to rage, he had struck and

[1] The speech to which this refers is the square-bracketed (optional?) deletion: see above, p. 236, note 3.

[2] Doubly underlined. 'It is [in the murder-scene], of course, that the excitement of the audience is wrought to its highest pitch, and that the acme of the actor's art is reached. The raised hand, the bent-back head, are good; but shut your eyes, and the illusion is more complete. Then the cries for mercy, the "Bill! dear Bill! for dear God's sake!" uttered in tones in which the agony of fear prevails over the earnestness of the prayer, the dead, dull voice as hope departs, are intensely real. When the pleading ceases, you open your eyes with relief, in time to see the impersonation of the murderer seizing a heavy club, and striking his victim to the ground' (Edmund Yates, op. cit., p. 63).

[3] *twice* doubly underlined.

[4] *Rose Maylie's* doubly underlined.

[5] *struck her* underlined doubly. Marginal stage-direction *Action*. Dickens wrote on 9 April 1869: 'I don't think a hand moved while I was doing [the murder] last night, or an eye looked away. And there was a fixed expression of horror of me, all over the theatre, which could not have been surpassed if I had been going to be hanged to that red velvet table' (*N*, iii. 718–19).—This ends ch. 47; *Berg* continues with the first paragraph of ch. 48, subsequently deleted by Dickens, and has against it the marginal stage-direction *Mystery*. (In *Suzannet* this is placed half-way down the paragraph.) In both texts, most of the underlining here is of the double interrupted kind.

[6] *the room*, and *now* below, doubly underlined.

[7] Marginal stage-direction *Terror to the End*; in *Berg*, this is emphatically capitalized—*Terror To The End*.

struck again. Once he threw a rug over it; but it was worse to *fancy* the *eyes*,[1] and imagine them moving towards him, than to see them glaring upward, as if *watching the reflection of the pool of gore that quivered and danced in the sunlight on the ceiling*. He had plucked it off again. And there was the body—mere flesh and blood, no more—but *such* flesh, *and so much blood*!!!

He struck a light, kindled a fire, and thrust the club into it. There was hair upon the end, which shrunk into a light cinder, and whirled up the chimney. Even that frightened him; but he held the weapon till it broke, and then piled it on the coals to burn away, and smoulder into ashes. He washed himself, and rubbed his clothes; there were spots upon them that would not be removed, but he cut the pieces out, and burnt them. *How those stains were dispersed about the room*! *The very feet of his dog were bloody*!!!![2]

All this time he had, *never once*, turned his *back* upon the *corpse*. He now moved, *backward*, towards the door: dragging the dog with him, shut the door softly, locked it, took the key, and left the house.[3]

As he gradually left the town behind him all that day, and plunged that night into the solitude and darkness of the country, he was *haunted by that ghastly figure following at his heels*. He could hear its garments rustle in the leaves; and every breath of wind came laden with that last low cry. If *he* stopped, *it* stopped.[4] If *he ran*, *it followed*; not running too—that would have been a relief—but borne on one slow melancholy air that never rose or fell.

At times, he turned to beat this phantom off, though it should look him dead; but the hair rose on his head, and his blood stood still, for it had turned with him, and was behind him then.[5] He leaned his back against a bank, and felt that it stood above him, visibly out against the cold night sky. He threw himself on his back upon the road. *At his head it stood, silent, erect, and still: a human gravestone with its epitaph in Blood*!![6] *

Suddenly, towards daybreak, he took the desperate resolution of going back to London. 'There's somebody to speak to there, at all events. A hiding-place, too, in our gang's old house in Jacob's Island.—I'll risk it.'

Choosing the least frequented roads for his journey back, he resolved

[1] *fancy* and *eyes*, and *so much blood* below, doubly underlined.

[2] *The very feet of his dog were bloody* trebly underlined. Vertical lines in the margin, against this sentence.

[3] Here the printed text of *Berg* (the version performed at the Trial Reading) ends. The extra narrative resumes at a point several pages later in ch. 48.

[4] The *he* . . . *it* . . . *he* . . . *it* were underlined in the *Berg* manuscript, and printed as italic in *Suzannet*, where they are all underlined; *ran* and *followed* are doubly underlined.

[5] In *Berg*, the words from 'but the hair' to 'behind him then' are underlined.

[6] *Blood* underlined doubly.

to lie concealed within a short distance of the city until it was dark night again, and then proceed to his destination. * He did this, and limped in among three affrighted fellow-thieves, the ghost of himself—blanched face, sunken eyes, hollow cheeks—*his dog at his heels covered with mud, lame, half blind, crawling as if those stains had poisoned him*!![1]

All three men shrank away. Not one of them spake.

'You that keep this house.—Do you mean to sell me, or to let me lie here 'till the hunt is over?'

'You may stop if you think it safe.[2] * But what man ever escaped the men who are after you!'

Hark!!!![3] A great sound coming on like a rushing fire! What? *Tracked so soon*? The hunt was up already? Lights gleaming below, voices in loud and earnest talk, hurried tramp of footsteps on the wooden bridges over Folly Ditch, * a beating on the heavy door and window-shutters of the house, * a waving crowd in the outer darkness like a field of corn moved by an angry storm!

'The tide was in, as I come up. Give me a rope. I may drop from the top of the house, at the back into the Folly Ditch, and clear off that way, or be stifled. *Give me a rope*!'

No one stirred. They pointed to where they kept such things, and the murderer hurried with a strong cord to the housetop. *Of all the terrific yells* that ever fell on *mortal ears*, *none could exceed* the furious cry when *he was seen*.[4] Some shouted to those who were nearest, to set the house on fire; others adjured the officers to shoot him dead; others, with execrations, clutched and tore at him in the empty air; some called for ladders, some for sledge-hammers; some ran with torches to and fro, to seek them. * '*I promise Fifty Pounds*,' cried Mr. *Brownlow*[5] from the nearest bridge, 'to *the man who takes that murderer alive*!' *

He set his foot against the stack of chimneys, fastened one end of the rope firmly round it, and with the other made a strong running noose by the aid of his hands and teeth. With the cord round his back, he could let himself down to within a less distance of the ground than his own height, and had his knife ready in his hand to cut the cord, and drop.

At the instant that he brought the loop over his head before slipping it beneath his arm-pits, *looking behind him* on the *roof* he *threw up his arms, and yelled*, '*The eyes again*!'[6] Staggering as if struck by lightning, he lost his balance and tumbled over the parapet. The noose was at his neck;

[1] Doubly underlined from *lame* to *poisoned him*. The final phrase ('crawling as if . . .') does not appear in the novel.

[2] On the textual differences from the novel, in the following sentences, see the head-note.

[3] *Hark* trebly underlined. Marginal stage-direction *Action*.

[4] *he was seen* doubly underlined.

[5] *Brownlow* doubly underlined.

[6] *yelled* and *again* trebly underlined; *eyes* underlined four times.

it ran up with his weight; tight as a bowstring, and swift as the arrow it speeds. He fell five-and-thirty feet, and hung with his open *knife clenched in his stiffening hand*!!!

The *dog* which had lain concealed 'till now, ran backwards and forwards on the parapet with a dismal howl, and, collecting himself for a spring, jumped for the *dead man's shoulders*. Missing his aim, he fell into the ditch, turning over as he went, and striking against a stone, *dashed out his brains*!![1]

THE END OF THE READING

[1] *dashed out his brains* underlined doubly.